TAGORE, EINSTEIN AND THE NATURE OF REALITY

This volume consists of a selection of scholarly essays from literature, philosophy and history on the conception of reality as understood by Rabindranath Tagore and Albert Einstein.

The nature of reality has been a long-debated issue among scientists and philosophers. Tagore (1861–1941) met Einstein (1879–1955) at the latter's house in Kaputh, Germany on 14 July 1930 and had a long conversation on this issue. This conversation has been widely quoted and discussed by scientists, philosophers and scholars from the literary world. The important question that Tagore and Einstein discussed was whether the world is a unity dependent on humanity, or the world is a reality independent of the human factor. Einstein believed that reality is independent of the mind and the human factor. On the other hand, Tagore adopted the opposite view. Nevertheless, both Einstein and Tagore claimed to be realists – their conceptions of reality were obviously fundamentally different. Where does the difference lie? Can it be harmonized at a deeper level?

This volume brings together for the first time a gamut of views on this subject from eminent scholars. It presents some key reflections on reality, language, poetry, truth, science, personality, human sciences, virtue ethics, intelligibility and creativity. It will be useful to scholars and researchers of philosophy, literature, history and political studies, as well as to those interested in Tagore.

Partha Ghose is a theoretical physicist and Honorary Scientist at the National Academy of Sciences, India. He retired as Professor, S. N. Bose National Centre for Basic Sciences, Kolkata in 1999. He is a Fellow of the National Academy of Sciences, India; the West Bengal Academy of Science & Technology; a member of the Board of Trustees of the Academy of Fine Arts, Kolkata; and former Chairman of the Satyajit Ray Film & Television Institute, Kolkata. He specializes in the foundations of quantum mechanics and unified theories. He was awarded the Indira Gandhi Prize for the popularization of science by the Indian National Science Academy. His exposition of Rabindranath Tagore's philosophy and music has found expression in several scholarly papers. His publications include the book *Testing Quantum Mechanics on New Ground* (1999).

TAGORE, EINSTEIN AND THE NATURE OF REALITY

Literary and Philosophical Reflections

Edited by Partha Ghose

Routledge
Taylor & Francis Group
LONDON AND NEW YORK

First published 2019
by Routledge
2 Park Square, Milton Park, Abingdon, Oxon OX14 4RN

and by Routledge
605 Third Avenue, New York, NY 10017

First issued in paperback 2020

Routledge is an imprint of the Taylor & Francis Group, an informa business

British Library Cataloguing-in-Publication Data
A catalogue record for this book is available from the British Library

Library of Congress Cataloging-in-Publication Data
A catalog record for this book has been requested

ISBN 13: 978-0-367-73133-5 (pbk)
ISBN 13: 978-0-8153-7211-0 (hbk)

Typeset in Sabon
by Apex CoVantage, LLC

CONTENTS

CONTRIBUTORS

Sourin Bhattacharya is former Professor of Economics at Jadavpur University, Kolkata, India. His areas of interest are theory, thought and method. He has written mostly in Bengali on demand theory, language of change, crisis of socialism, development, the politics and culture of transparency, Tagore and modernity. He has translated and edited selected essays of Gramsci jointly with Samik Bandyopadhyay.

Bijoy Boruah retired recently from the Indian Institute of Technology Delhi as Professor of Philosophy. Previously, he taught at IIT Kanpur for 21 years. Currently, he is Visiting Professor in the Department of Humanities and Social Sciences in the Indian Institute of Technology Ropar. His research interest is focused on the metaphysics of the self.

Sitansu Sekhar Chakravarti is Affiliated Visiting Scholar at New College, University of Toronto, Canada. He has published on the philosophy of language of the West, ethics in the Mahabharata, and Hindu spirituality, combining Indian and Western philosophical perspectives. He has been working on virtue ethics, and the philosophy of Rabindranath Tagore for several years.

Supriya Chaudhuri is Professor (Emeritus) in the Department of English, Jadavpur University, Kolkata, India. Her specializations include European Renaissance literature, Bengali cultural history, modernism, narrative and theory. She has translated Bengali fiction and poetry, including the writings of Rabindranath Tagore for the *Oxford Tagore Translations* series and *The Essential Tagore* (2014).

Probal Dasgupta is former Professor at the Indian Statistical Institute, Kolkata, India. He has taught in New York, San Francisco, Melbourne,

Barlaston, Kolkata, Pune, Hyderabad, and is an Honorary member of the Linguistic Society of America (since 2004). He is President of the Akademio de Esperanto 2016–2019 and was President of the World Esperanto Association, 2007–2013. His books in English include *The Otherness of English: India's Auntie Tongue Syndrome* (1993) and *Inhabiting Human Languages: The Substantivist Visualization*. He has over 500 Bangla/English/Esperanto publications in linguistics, literature, translation and philosophy.

Amiya Dev has taught Comparative Literature at Jadavpur University, Kolkata, India. He has written widely in Bengali and English on Tagore and after, his latest Tagore book being *Rereading Tagore* (2018). He has also edited a few volumes including *Science, Literature and Aesthetics* (2009) for the Centre for Studies in Civilizations, and co-edited such others as *Tagore and China* (2011).

Ashish Lahiri is a writer, translator, lexicographer and an independent researcher in the History of Science, specializing in the science – culture interface. Well-known for his Bengali translation of J. D. Bernal's magnum opus *Science in History*, Lahiri is currently a Visiting Faculty member, History and Philosophy of Science, Indian Institute of Science Education & Research (IISER), Kolkata, India.

Bimal Krishna Matilal (1935–1991), Spalding Professor of Eastern Religion and Ethics at the University of Oxford (1977–1991), was an Indian philosopher whose influential writings present the Indian philosophical tradition as a comprehensive system of logic. His works helped create a vibrant revival of interest in Indian philosophical tradition as an important source of ideas.

Shefali Moitra is former Professor of Philosophy and former Director, School of Women's Studies, Jadavpur University, Kolkata, India. She has published several articles on Tagore in journals and anthologies, both in English and in Bengali; these include a detailed feminist reading of Tagore's Dance Dramas. Her areas of interest are feminist thought, ethics and philosophy of language.

Nirmalangshu Mukherji is former Professor of Philosophy, University of Delhi, India and National Professor of ICPR in 2015–2016. His recent books include *The Primacy of Grammar* (2010), *Reflections on Human Inquiry* (2017) and *The Structure of the Human Mind* (forthcoming).

Syed Sayeed is Professor of Philosophy at the English and Foreign Languages University, Hyderabad, India. After obtaining his doctorate from IIT Kanpur, he taught at the Aligarh Muslim University for several years. He was at Liverpool University as a Charles Wallace Trust Visiting Fellow. He has published a book and several research articles in journals of repute on a variety of subjects such as philosophy of social sciences, literature and political philosophy.

Saranindranath Tagore is Associate Professor in the Philosophy Department, National University of Singapore. His main teaching and research interests are in the areas of twentieth-century Continental/European Philosophy (French and German traditions) and Indian Philosophy, especially the modern tradition. He has been a visiting scholar at the Munk Center of International Studies at the University of Toronto and at the Harvard-Yenching Institute and was appointed an Affiliate Fellow, South Asia Initiative at Harvard University (2009–2010). He was conferred an honorary professorship by Zhejiang Yuexiu University of Foreign Languages (China, 2016). He is one of the Editors-in-Chief of *Sophia: International Journal of Philosophy and Tradition*. Apart from his philosophical publications, he is also a literary translator; his work includes *Rabindranath Tagore: Final Poems* (2001).

INTRODUCTION

Partha Ghose

Rabindranath Tagore and Albert Einstein, two of the most iconic figures of the twentieth century, met at least six times. It might seem as though a poet and a scientist hold little in common. But Einstein and Tagore shared many things. Tagore received the Nobel Prize in Literature in 1913 '*because of his profoundly sensitive, fresh and beautiful verse, by which, with consummate skill, he has made his poetic thought, expressed in his own English words, a part of the literature of the West.*' Eight years later in 1921 Einstein was awarded the Nobel Prize in Physics '*for his services to Theoretical Physics, and especially for his discovery of the law of the photoelectric effect.*' They lived at a time when profound social, political and historical changes were sweeping across the world. Both of them publicly voiced concern over the rise of aggressive nationalism, and upheld the cause of human rights and creative freedom in the pursuit of world peace. At the invitation of Romain Rolland (a Nobel Laureate in Literature, 1915), they signed an anti-nationalist document, 'declaration pour l'independence de l'esprit' in 1919. Other signatories included Jane Addams, Tolstoy's secretary Pavel Birukov, Benedetto Croce, Georges Duhamel, Hermann Hesse, Selma Lagerlöf and Bertrand Russell, among others.

Tagore's first visit to Germany was in 1921. In Martin Kämpchen's words,

> Tagore mesmerized and fascinated his German audiences. Wherever he spoke, the halls were packed. Indeed, the newspapers reported scuffles and regular fights by people who were refused entry. The German press rose to the occasion by reporting Tagore's every movement. . . .
>
> For Europeans, Tagore represents the distant memory of a Wise Old Man from the East, of an Eastern mystic who

> arrived in Germany after the First World War to dispense consolation and courage to a people immersed in a deep spiritual and cultural crisis. For less than a decade, Tagore enthused German audiences and readers, after which he sank into oblivion, a process which was aided by the advent of Nazi Germany for which the Indian poet was anathema.[1]

Many European intellectuals of the 1920s including Rainer Maria Rilke, Stefan Zweig, Count Hermann Keyserling, Herman Hesse, Andre Gide, Romain Rolland, Sylvan Levy, Ludwig Wittgenstein and scientists like Einstein, Arnold Sommerfeld, Max Born and Werner Heisenberg admired Tagore. His book of poems *Gitanjali* and plays like *Home and the World* were translated into German and many European languages.[2]

He visited Germany again in 1926 and 1930. Just before his 1930 visit, Tagore delivered the Hibbert lectures in Oxford, which were later published as *The Religion of Man* in which Tagore expounded his philosophy of the humanity of God and the divinity of Man. Soon after these lectures he went to Germany and met Einstein. The first conversation on the nature of reality was reported in *The New York Times* on 10 August 1930 with the heading *Einstein and Tagore Plumb the Truth*. The reporter was Einstein's step son-in-law, Dmitri Marianoff who, together with Amiya Chakravarty who accompanied Tagore and who was a poet in his own right, had taken notes which were later approved by both Einstein and Tagore. There exist slightly different versions of the conversation, such as in Tagore's *The Religion of Man* (Appendix II, 1931 edition), the typescripts in the archive at Rabindra Bhavana, Santiniketan and *The Kenyon Review*.[3] Here is Marianoff's report:

> Near Potsdam is a small place called Caputh. There, upon a hill, stands a brown wooden house with a red tile roof. . . . At about 4 o'clock one recent afternoon Rabindranath Tagore walked along the sandy path to the house. . . .
>
> Atop the hill, Tagore sat down in an armchair on the lawn with Einstein and his family to enjoy the scene. . . .
>
> A lively discussion arose. It was interesting to see them together – Tagore, the poet with the head of a thinker, and Einstein, the thinker with the head of a poet. . . .
>
> They simply exchanged ideas. But it seemed to an observer as though two planets were engaged in a chat.

An impression has gained ground that the Einstein-Tagore conversation was a failure, perhaps because of Marianoff's comment, 'it seemed to an observer as though two planets were engaged in a chat.' In this context the famous Bohr-Einstein debate comes to mind. Unlike a brief afternoon chat between a scientist and a poet through an interpreter, the Bohr-Einstein debate was between two of the leading scientists of the time on a subject which both helped to create and develop, and it was extended over nearly 28 years from 1927 to Einstein's death in 1955. The contention was essentially the same, namely the role of an observer in the description of nature, though couched in the technical language of physics. Bohr kept on pressing his point about the essential role of measurements by observers in describing nature, while Einstein steadfastly stuck to his faith in an observer-independent reality. They did not budge from their respective positions, while maintaining dignity and commitment to logical thinking. Einstein was obviously disturbed by the fact that the vast majority of physicists shared Bohr's point of view. Perhaps it was this concern that made him raise the question with Tagore whose views he valued.

Einstein was initially enthusiastic about having the transcript of the conversation approved by Tagore and published. Later, for some reason, he seemed dissatisfied with it and wished it had never been published. Tagore himself wrote an account of his meetings with Einstein entitled *My Memories of Einstein* (reproduced here as Appendix 1) in which he says,

> Between us was a sympathy which only the barriers of language made awkward. His mastery of English is not enough for conversation and I cannot speak German. The interpreter between us must have had a not too easy task.

Despite all these reservations about the success of the conversation, it has been commented on by many eminent scientists and philosophers in recent times including Brian Josephson (Nobel Laureate in Physics, 1973), Ilya Prigogine (Nobel Laureate in Chemistry, 1977), Abraham Pais, physicist and scientific biographer of Einstein and Niels Bohr, and J. N. Mohanty, the philosopher.

According to Josephson,

> Einstein was not understanding what Tagore was saying. The point at issue was whether we have an objective world or a human world. Einstein thought that there was objective reality and that Tagore's position was nonsense. I think my

> comment at the time was that Einstein was not appreciating how much the process of construction, engaged in by our senses and minds, affect what we see and what our science consists of. . . . So, to summarize, I think one can certainly justify by a number of arguments Tagore's point that our science is a human activity and our universe also may be a human universe, possibly something that came into existence at the time of the big bang. . . .
>
> Tagore is, I think, saying that truth is a subtler concept than Einstein realizes.[4]

Ilya Prigogine has a slightly different perspective:

> Einstein emphasized that the science had to be independent of the existence of any observer. This led him to deny the reality of time as irreversibility, as evolution. On the contrary, Tagore maintained that even if absolute truth could exist, it would be inaccessible to the human mind. Curiously enough, the present evolution of science is running in the direction stated by the great Indian poet. Whatever we call reality, it is revealed to us only through the active construction in which we participate.[5]

What keeps it alive even today?

To examine its contemporary relevance a three-day International Seminar on 'The Nature of Reality: The Perennial Debate' was organized from 1 to 3 March 2012 at the Indian Institute for Advanced Study, Shimla with the intention of celebrating Tagore's 150th birth anniversary with wide-ranging discussions on the entire gamut of man's engagement with reality through science, philosophy, language, cognitive neuroscience, literature and the history of science. Some of these lectures dealing with the philosophy of science, supplemented by a few others that were not presented at the seminar, were published as *Einstein, Tagore and the Nature of Reality* by Routledge, Taylor & Francis Group, London and New York, in 2017. The present volume is a supplement to it, consisting as it does of a selection of the lectures delivered at the seminar from the points of view of literature and general philosophy, again supplemented by a few other contributions that were not presented at the seminar but are felt to be relevant to the issue.

A proper understanding of the famous conversation requires acquaintance with the philosophical predilections of each of the outstanding personalities. This is a daunting task for a poet like Tagore who was so prolific in his writings and had such an informal but very original style of doing philosophy, distinctly his own, with which very few are familiar. The same cannot be said of Einstein, most of whose works lie squarely within the realm of theoretical physics, and whose views on reality are relatively easier to understand. Hence it is but natural and appropriate that more would be said about Tagore than Einstein in a volume on literary and philosophical dimensions of the concept of reality. In fact, there is a need for some chapters to be entirely devoted to Tagore, for it is only in that way that the reader will come to appreciate the subtlety and nuances of his profound views as compared to the more directly understandable scientific view that Einstein took. In particular, his views of aesthetic and literary creation are extremely important in this context. This is the very *raison d'être* of this book, the philosophy of science having already been covered by the companion volume *Einstein, Tagore and the Nature of Reality* referred to above.

The book brings together for the first time a variety of scholarly analyses of Einstein's and Tagore's views on the nature of reality from the standpoints of general philosophy, literature and linguistics. All the authors have dealt mainly with Tagore's human conception of reality as opposed to Einstein's human-independent view. Few seem to be aware that Einstein himself championed the role of creative human imagination in scientific theory building, a point which even Brian Josephson seems to have missed. Let me quote a few lines in support of this so that a more comprehensive picture emerges of Einstein's views. In his *Autobiographical Notes* Einstein writes:

> The prejudice . . . consists in the faith that facts by themselves can and should yield scientific knowledge without free conceptual construction. Such a misconception is possible only because one does not easily become aware of the free choice of such concepts, which, through verification and long usage, appear to be immediately connected with the empirical world.[6]
>
> A theory can be tested by experience, but there is no way from experience to the setting up of a theory.[7]

To see how strongly he believed in the crucial role that theory plays in unravelling 'scientific truth,' let us now see how Heisenberg recounts his first encounter with Einstein that led to his discovery of the famous 'uncertainty principle', which flew against all established scientific principles:

> For the first time, therefore, I now had the opportunity to talk with Einstein himself. On the way home, he questioned me about my background, my studies with Sommerfeld. But on arrival, he at once began with a central question about the philosophical foundation of the new quantum mechanics. . . . He thought that every theory in fact contains unobservable quantities. . . . He pointed out to me that the very concept of observation was itself already problematic. . . . In that case, *it is the theory which first determines what can be observed.*[8]
> (italics added)

Why then did he appear to question Tagore's grounding of reality in the human mind? Could it be because he saw in physics a certain reliability and certainty that he could not find in human affairs? Or could it be that he saw in Tagore's view a justification of Bohr's standpoint and that of most contemporary physicists, which was anathema to him and against which he was hoping to receive Tagore's support? Is this why he was unhappy that the conversation was published?

Einstein was keenly aware of the inadequacy of the scientific method when it came to questions of ethics. In an address before the Princeton Theological Society in May, 1939 this is what he had to say:

> the scientific method can teach us nothing else beyond how facts are related to, and conditioned by, each other. . . . Yet it is equally clear that knowledge of what *is* does not open the door directly to what *should be*. . . .
>
> If we ask whence derives the authority of such fundamental ends, since they cannot be stated and justified merely by reason, one can only answer, they exist in a healthy society as powerful traditions, which act upon the conduct and aspirations and judgements of the individuals; they are there, that is, as something living, without its being necessary to find justification for their existence. They come into being not through demonstration but through revelation, through the medium of powerful personalities. One must not attempt to justify them, but rather to sense their nature simply and clearly.[9]

With that background let me now briefly summarize the 12 essays in this volume.

In the Preamble, B. K. Matilal develops Tagore's man-centric view of the universe in response to Einstein's query at their meeting in 1930. Science may question it, holding on to the materiality of the world, but Matilal makes the point with characteristic clarity and brevity that there can be no knowledge, even of the material world, without human intervention. For all knowledge is held in a language, and language is a medium made by man. Hence no truth is utterly independent of man.

In the following chapter Nirmalangshu Mukherji begins by pointing out Tagore's obvious point that physics itself is a human enterprise just like any other human creation such as music, poetry, painting and sculpture. He then asks: what effect does this alone have on the issue of realism? He argues that the "humanness" of physics cannot by itself lead to anti-realism about entities and theories proposed in physics; otherwise realism cannot even be formulated. It follows then that the realism-antirealism options must be framed within the suggested "humanness." A formal way to conceptualize the problem would be to examine Immanuel Kant's proposal that a sense of (independent) reality is itself a part of the human cognitive apparatus: the "Galilean Style" is a product of this apparatus.

In the second chapter Probal Dasgupta dwells on a less known aspect of Tagore's contestation of the 'scientific' analysis of language. After Suniti Kumar Chatterji's magnum opus *The Origin and Development of the Bengali Language* (1926) appeared, Tagore and Chatterji spent several months together in Indonesia. Their friendship was soon enlivened by articulate disagreement. In 1938 Tagore contested Chatterji's historical science of language by his own description of the Bengali language, which Dasgupta construes as a 'methodological reprise and a philosophical upscaling of his critique of Einstein.' Such a construal may help readers of this volume who are familiar with the Chomsky revolution which has made mainstream linguistics – against whose foundations Tagore's critique is directed – forge intimate bonds with the methodology of physics itself.

In the third chapter Amiya Dev approaches the issue of the nature of reality from the point of view of literature. To Dev, Tagore is primarily a creative artist. His poems and songs, his plays, his musical and dance dramas, stories and novels, essays and discourses, autobiographies and travelogues, even his letters are where his 'truth' abounds, more than in his 'philosophical' writings in English. He develops his theme on a rereading in recent years of Tagore's literary works, and delves into Tagore's truth in them, paving the way to a better understanding of his dialogue with Einstein and of Tagore as a creative artist with human concerns that are at times at odds with the scientist's objective view of the world.

In the next chapter Supriya Chaudhuri begins by examining the literary techniques of realism as practiced and commented on by Tagore. Chaudhuri points out that in his fictional and non-fictional prose, no less than his poetry, Tagore offers a critique of the simple identification of 'the real' with 'the things' of the world. This may be seen in his short essay 'Bastab' [The Real], first published in the journal *Sabuj Patra* in 1914 and later reprinted as the opening chapter of the volume *Sahityer Pathe* (1936), where he appears to argue that 'the real' is an ideological construction, not a given of human perception. Chaudhuri draws examples from Tagore's prose writings in order to investigate a literary debate on 'realism,' rather than a metaphysical questioning of 'the nature of reality.' In the concluding section, she turns to the disagreements between Einstein and Tagore in their conversations of July 1930. These disagreements, she argues, inhere in a difference of philosophical ground: first of all in different understandings of the term 'human,' second, in divergent conceptions of consciousness, and finally, on the impossibility of reconciling the physical with the metaphysical.

In Chapter 5 Shefali Moitra begins by noting that Einstein's conceptual scheme could not have been worlds apart from that of Tagore. There is obviously no difference between Einstein and Tagore that a scientist aims at being detached and unbiased in his investigation of the world. But Tagore keeps shifting between the domains of metaphysics and epistemology. According to him, the infinite being expresses himself through the medium of finitude, the world. The finite world is only the medium through which the infinite is expressed. Science is reluctant to accept the paradox of the infinite assuming finitude. Tagore insists the paradox cannot be wished away. According to Moitra Einstein took Tagore to believe, if there is no human mind, then there is no world either, but that was not Tagore's contention. What he held is, if there is no human mind, then there is no humanly known world. When Tagore deals with the broad outline of science, he comes very close to Einstein. When Tagore deals with religion, he moves worlds apart. Drawing from Tagore's *Sadhana, Personality* and *The Religion of Man* Moitra attempts to highlight the points where Tagore comes close to Einstein and where he moves away.

In Chapter 6 Sourin Bhattacharya looks at the question of reality and raises several questions: What is the nature of reality? What should be our attitude towards it? Is it totally independent of man like something out there? Or is it a Man's universe? Einstein and Tagore differed sharply though politely. They were perhaps talking only about the natural sciences. In this chapter Bhattacharya widens the scope of the discourse to include also the human sciences. He briefly goes through the 'sciences war' perspective of the 1990s and its earlier version of the 'two cultures' debate of the 1960s. This is intended to bring the two iconic minds closer but with an acknowledgement of technical barriers to a fuller understanding of the issues involved. Even then, Bhattacharya believes it is culturally important that the dialogues should go on. Finally, he proposes a comprehensive way of approaching the nature of reality in terms of a relation space. Our concrete situation in the natural as well as the social world may be conceptualized as structured by networks of relations, which we often impute in the particular contexts of given theories. Thus reality, as we handle it, becomes imputed reality. There may be a transcendental layer of the reality that we may never comprehend.

In Chapter 7 Saranindranath Tagore offers a phenomenological reading of the idea of the human which Tagore presents to Einstein. He makes three interrelated points. First, he develops the idea of the human as a limit-concept with the aid of phenomenology as it emerges in the context of the conversation. Second, he tries to connect the themes of the conversation to Tagore's social philosophy. In the process he tries to establish the point that the idea of the human is grounded in cosmopolitanism. Finally, he arrives at the conclusion that Tagore was not attacking realism (as Einstein thought) but rather systematically extending his richly worked out cosmopolitan concerns.

In Chapter 8 S. Chakravarti deals with Tagore's concept of reality in the backdrop of Wittgenstein's *Tractatus Logico-philosophicus*. Wittgenstein was fond of the poet's writings, especially his play *The King of the Dark Chamber*. Chakravarty discusses Wittgenstein's preoccupation with the place of value in humans against the backdrop of scientific study and analysis, the object of which he technically designates as *the world* in the *Tractatus* and *which* does not incorporate value. One finds a secular concept of spirituality in the mysticism of the *Tractatus* that logically transcends the world of science, without positing any metaphysical entity as a basis for value. Chakravarty points to some philosophical reasons that might account for the strong attraction the philosopher had for the poet's play, even as he highlights some important differences in their respective positions. To the poet value is phenomenologically present without demanding any ontological mooring, which accounts for his secular spirituality.

Tagore's thoroughgoing phenomenological perspective makes his world a human world. The virtue of *harmony within* is pivotal to his virtue ethics, and his activism moves toward building harmony where it is found missing. His existentialistic ideas flourished when existentialism had not yet become a prominent movement in the European continent. The article adds a new dimension to *Tractatus* studies by helping the understanding of mysticism in it. Chakravarty avoids the use of technical expressions, rendering his exposition accessible to a non-philosophical reader.

In Chapter 9 Syed Sayeed analyses the dialogue from a philosophical perspective. He elaborates and makes explicit the two distinct conceptions of reality present in the worldview of each. Tagore bases the existence of the world on human truths, while Einstein perceives a universal, impersonal reality. Sayeed tries to bring out the nuances and timeless relevance of each.

In Chapter 10 Bijoy Boruah deals with the relation between reality and truth as problematized in philosophy in relation to science as well as the humanities. Metaphysics – the Theory of Reality – commands independence from epistemology – the Theory of Knowledge; what is (metaphysically) real is so irrespective of whether we know it or not, or how we claim to know it. But, Boruah points out that to afford a metaphysical *conception* of reality is to concede that ultimate reality is *intelligible* to the human mind or consciousness. Einstein once famously wrote, 'The eternal mystery of the world is its comprehensibility. . . . The fact that it is comprehensible is a miracle.' The problem about 'reality and truth' is therefore that of formulating the idea of 'metaphysical intelligibility.' And the crux of this formulation is the requirement that the intelligence in question has the right measure of objective adequacy. It would then be a conception of an ideal intelligence (mind or consciousness) that is intrinsically capable of transcending all limitations or deficiencies immanent to any perspectival representation of reality. Indeed, the aspiration for an aperspectival conception of reality is central to the metaphysical project of relating truth with reality. This is the Einsteinian credo. However, the metaphysical project cannot be insensitive to the perspectival character of 'lived reality,' because the subjective reality of the life-world is part of the overall conception of reality. Since consciousness is part of the fabric of reality, conscious existence is necessarily a 'lived' existence, which is lived subjectively, given the nature of consciousness. A theory of reality is therefore supposed to extend its scope to take both perspectival and aperspectival conceptions into its comprehensive fold. This is the Tagorean credo.

In the final chapter Ashish Lahiri deals with Tagore's attempt to edit the *Upanishads* to suit his modern science-honed perception of reality. He does not dismiss the difference between 'scientific truth' and what he calls the 'universal truth.' Scientific truth 'can only be reached through the process of logic,' but that very logic is 'human'; hence scientific truth is dependent on the human entity. This position may be characterized as 'cosmic scepticism.' Yet, Tagore was unequivocal in emphasizing that modern culture must imbibe theoretical science as an inalienable part of its *aesthetics*. In his 1914 essay *Āmār Jagat* (*My World*), he says the mind is not merely a mirror that reflects reality, but a component (*upakarana*) of that reality. He describes a mock debate between a poet and his scientist friend. It represents a water-shed in Tagore's ideas so far as the problems of subjectivity of art and the so-called objectivity of reductionist science are concerned. The two major components of Tagore's mind vis-à-vis science are: 'making playthings of facts' fished out by scientists and deriving 'pleasure,' even 'thrill,' from the 'analytical view of objective reality.' As an example of Tagore's cosmic scepticism sublimated into high poetry, Lahiri suggests one should consider the famous song *vānī mor nāhi*, composed in 1939, nearly at the end of his life.

Notes

1 Martin Kämpchen, *Rabindranath Tagore in Germany*, Parabaas, July 25, 2003, https://www.parabaas.com/.
2 For a contemporary assessment of Tagore from the point of view of the West, see Adam Kirsch, 'Modern Magus: What did the West see in Rabindranath Tagore?' *New Yorker*, May 30, 2011.
3 *The Kenyon Review* (New Series, vol. 23, no. 2, *Cultures of Creativity: The Centennial Celebration of the Nobel Prizes* (Spring, 2001), pp. 20–5: published by Kenyon College.
4 Dipankar Chattopadhyay, *Rabindranath O Vijnan*, Ananda Publisher, 2000.
5 Ilya Prigogine and Isabelle Stengers, *Order Out of Chaos*: *Man's New Dialogue with Nature*, London: Fontana Paperbacks, 1985, p. 293.
6 A. Einstein, 'Autobiographical Notes', in P. A. Schilpp (ed), *Albert Einstein: Philosopher-Scientist*, Evanston, IL: Library of Living Philosophers, 1949, p. 49.
7 Ibid., p. 89.
8 W. Heisenberg, *Encounters with Einstein*, Princeton, NJ: Princeton University Press, 1989.
9 A. Pais, *Einstein Lived Here*, Oxford and New York: Clarendon Press, Oxford University Press, 1994, pp. 120–1.

PREAMBLE

A man-centric view: Poet and scientist

Bimal Krishna Matilal
Translated from Bengali by Amiya Dev

Exactly 60 [82]* years ago Tagore put up his views in the Hibbert Lectures at Oxford. The subject was *The Religion of Man*. These lectures given in 1930 were published the following February. The publisher was Oxford University Press. The lectures had along with them an introduction by the poet and a few appendices. Recently the book has been reissued in soft covers. It has been edited by Andrew Robinson.

Here an introduction by the editor has been added, though the appendices of the earlier edition have been removed – the book seems to have been maimed by that. Among these appendices was a description of the poet's meeting with the world-famous scientist Albert Einstein. At the outset of my discourse I would like to give a quotation from it. From a concise summary of that we are likely to get an idea of Tagore's religious and spiritual thoughts.

In answer to the scientist's query Tagore expressed his views on the universe and the truth. He said – 'When our universe is in harmony with Man, the eternal, we know it as truth, we feel it as beauty.' The scientist instantly said – 'This is a purely human conception of the universe.' This man-centric view of the universe seems to me to be a basic thrust of the poet's religious and spiritual thoughts. We may overlook this as the poet's fancy or may deign to accept it as a 'poetic truth,' but many modern rational men may hesitate to take it as a philosophical truth. Tagore was aware of this view of rational men's or of their objection. So he had a ready reply to that and had at various times given that in various ways. In the poem 'Ami' (I) included in the book *Syamali* he said –

> You will say, this is theory,
> not a poet's words.

> I shall say, this is truth,
> hence poetry.
> This is my pride,
> pride on behalf of all mankind.
> On the very canvas of man's pride
> is the great creator's art, the universe.

In the last two lines is contained a truth belonging to his view of life. So with great confidence did Tagore answer Einstein (apropos Einstein's comment mentioned above):

> There can be no other conception (no other than 'a man-centric view of the universe'). This world is a human world – the scientific view of it is also that of the scientific man.

Here in simple words and briefly a rather sound argument has been put forward. The scientist is trying to see the material world in such a way as contains no colour of human imagination – the directly perceived world is taken as that world by us, the world unique. Hence the direct evidence has been placed foremost in gaining knowledge about the world. The scientist also takes the direct evidence and assumptions based on the direct evidence as the only dependable means for his search for truth. But there is serious doubt among philosophers as to if a unique knowledge of the material world is at all possible. The simple argument against that is the following – knowledge has to be expressed in a language, else that knowledge cannot be communicated to others. And the language that is the medium of knowledge is created by man – in different languages are a number of major philosophical truths reflected. Therefore, to express that which we regard as material bereft of imagination we shall have to take recourse to a man-made medium, yet from that medium human imagination cannot be totally banished. When we call a thing sweet it is our tongue that tells us so. A specific kind of taste or feeling is called 'sweet.' Another kind of taste is called 'bitter.' But the same sweet tasting thing may prove bitter to a person suffering from bile – the very opposite. But the thing has not changed at all – it has remained the same in taste as before. It can be shown in this way that the material world cannot be at all described by totally leaving out the human attitude. Language carries the idea, yet that language is a medium made by man. Then how can we leave out man? This was the poet's question. So how else would we realize the world if not as a man-centric world? Besides what do we gain by that?

In the paragraph above I have placed a complex philosophical argument as simply as possible. But the point is, the argument cannot be summarily discarded – nor can this view be ignored and thrown out. Still our materialist mind does not want to agree that even in what we see with our eyes is the colour of imagination mixed invisibly to all. It is true that Einstein spoke from the scientist's standpoint of a scientific view of the universe – which is posited against the man-centric philosophy of life. We must admit that the philosophy of life to which scientists dedicate their life and through which our science and technology have made such advance, may not be dismissed as trifling. Tagore too did not want that. On the other hand he recognized it and said that it too contained hopes for mankind. Else the scientist's so-called scientific philosophy of life turns into a philosophy of life – colourless, meaningless, purposeless, intention-less, heartless, devoid of morality or moral sense, and devoid of the taste of beauty. Even though it may be true, it is unbearable. And probably it is not fully true either for we cannot altogether leave out man. For the search of truth, to establish the reality of matter, we have to abandon all foggy shades of imagination – lest we are misguided. But the truth too has a background – a perspective – how are we going to leave that out? And if we leave that out how are we going to find the truth – or how are we going to know it? If not reflected in the background, an image or a ray of 'light' is not seen by us – in that case how are we going to see the light of truth? It is the human mind that creates that background. Calling it 'imagination' may be belying the truth. In the poet's words this is man's 'mind-made grace.' It is by blending with this grace that we lay out the truth. If the scientist wants to use science for the good of mankind, then he too will have to go by that practice. Science gathers a lot of facts about a lot of things and gifts them to us – seen in a light free of imagination these facts are revealed to us. But the search afterwards for theories from the facts, the search for morals, for meaning, the follow-up of purpose – all that is the contribution of man-centric scientists and philosophers. Unless facts are judged in the light of theories, they are not properly valued.

The theory I am speaking of here is of world-wide validity – why the scientist alone, every man and woman seek this world-centric theory in their daily life – look for its significance – not that it has always to be the 'absolute' theory. Some people may go for the absolute theory, but not all – this has been indirectly endorsed by Sankaracarya as well. This world-wide theory is manifest as the 'poetic truth' in a novel manner. Aiming at it Tagore has said (in the 'I' poem) –

> Do not call it theory:
> My mind has become joyous
> in the creation of the 'Universal-I'
> with a brush in hand, and paint in a pot.

It is the philosophers that accede to the presence of a conscious self in all things of the world who will realize the true theory of this 'Universal-I.' Even though there may be an apparent conflict between materialist science and this man-centric view – they may be reconciled. It is a means thereof that is cited here.

Note

* The Bengali original ('Manabkendrik Darsan: Kabi o Bijnani') was reprinted in the Bengali fortnightly *Desh* in 2013 from the author's earlier essay on 'Tagore's Religious and Spiritual Thoughts' ('Rabindranather Dharma o Adhyatmacinta') published there in the Literature Number ('Sahitya Samkhya') in 1991, of which the present chapter was a part under the title 'Manabkendrik Darsan' (Man-Centric View).

1

POETIC REALITY

Nirmalangshu Mukherji

1

During his conversation with the physicist Albert Einstein, the poet-novelist-playwright-painter-composer-educationist Rabindranath Tagore stated that 'truth, which is one with the universal being, must be essentially human; otherwise, whatever we individuals realize as true, never can be called truth.' 'The truth which is described as scientific,' he continued, 'can (only) be reached through the process of logic' which itself is 'an organ of thought which is human.'[1]

The point of interest in these statements is that Tagore is concerned with truths of physics, scientific truth. For him, scientific truth is one with Universal being; in other words, truth is universal. While claiming that truth is 'human,' there is no emphasis on social norms, cultural forms, historicity, artistic variation, etc. In particular, he is not proposing that truth is subjective, whatever that problematic notion amounts to. In that sense, truth is objective; maybe *all* truths – not just truths of physics – are objective!! He seems to be upholding the concept of objective truth *while* asserting its essential humanness. In fact, he suggests that humanness is a *necessary* condition for objective truth! As we will see, this 'poetic' notion of truth lends a new dimension to the classical debate in philosophy of science between realism and anti-realism.

As noted, Tagore was a great poet and a humanist. As a humanist thinker, he also lectured and wrote on a variety of topics of general human interest such as the character of human existence, the play of nature on human creativity, and the role of values and religion in human societies. Such concerns abound in much of his creative work too, especially in his poems, plays and paintings. His Hibbert lectures at Oxford University in 1931, the Harvard lecture in 1913, the Kamala lecture at Calcutta University in 1933, and some other writings such as the collection of addresses given at the weekly prayer at Santiniketan, were more self-consciously and directly concerned with these topics.

In one broad common sense of the term 'philosophy,' the writings just mentioned may well be characterized as philosophical insofar as they are general reflections on the human condition. In Tagore's case, these reflections are often accompanied by short citations and commentaries from the classical Indian tradition such as the *Veda*s, *Upanishad*s, *Gita*, and the like, which also may be viewed as philosophical in roughly the same broad common sense of the term.

However, these reflective works are not philosophical in the narrower, more academic, sense. Broadly viewed as systematic reflections on the nature of language, thought and reality and the relations between them, academic philosophy can be safely identified, as with other academic disciplines, with its textual lineage.[2]

Consider the classical problem of realism opened by the Greeks, especially by Plato: does the world consist of the entities that correspond to the concepts humans employ to form a view of the world? In other words, is the human view of the world real? With the overwhelming fact of Newtonian science in hand, the problem took a particularly sharp form in the philosophy of Immanuel Kant, as we will see. Newer, more foundational, issues emerged as Newtonian physics gave way to relativity theory and quantum theory in early twentieth century. By the time the discussion between Tagore and Einstein took place in 1930, philosophy of science had already emerged as a significant area in philosophy with the realism/anti-realism debate at its core. There is no evidence that Tagore was familiar with the technical literature. To emphasize, there is no evidence that Tagore was familiar with the technical literature in the philosophy of science as developed in the works of, say, Ernst Mach, Henry Poincare and Karl Popper. To say this is not to deny that Tagore took intellectual – even if semi-popular – interest in aspects of modern science.

Tagore's views on physics and science, therefore, need to be viewed not as technical comments, but in terms of general humanist reflections of a literary mind. Even then Tagore's 'outsider' view carries much intellectual interest. A work of art, including the writing of poetry, is not merely a play of form. Somehow the artist has to relate the emerging forms in his aesthetic imagination to stable aspects of human experience for the forms to have an irresistible interpretation. In that sense, an artist constantly struggles with the elusive reality of artistic depictions. An artist and poet of Tagore's genius was likely to have reached a satisfactory reflective understanding of his own artistic expressions that might have interesting philosophical implications.

Elsewhere,[3] I have discussed how Tagore's poetic contemplation of bird-songs [*pakhire diyecho gaan*, *gay sei gaan* (*Balaka* 23)] illuminates an empirically viable conception of human musical ability. I wish to adopt a similar strategy for Tagore's 'poetic' views on science and truth.

Albert Einstein also was not an academic philosopher in the sense outlined above. However, his philosophical location with respect to the issue of scientific realism was far more intimate than Tagore's. Not only was Einstein the greatest physicist after Newton, his work was central in the new debate on scientific realism. In particular, he directly intervened in the post-quantum debate in scientific realism to suggest that quantum theory is an 'incomplete' theory of nature in that it fails to reach genuine (=objective) scientific truth. Thus, both for his outstanding role as a practicing scientist and a reflective thinker on the nature of science (specially, new physics), Einstein's remarks on the nature of reality carry intrinsic significance.

2

As we know by now, at the time of the conversations, Tagore was a very familiar name in Germany as some of his literary works had been translated into German and millions of copies were sold. Tagore visited Germany several times between the two world wars. During these visits, he met many leading German intellectuals such as Rainer Maria Rilke, Albert Schweitzer, Thomas Mann, Stefan Zweig, Hermann Hesse and others.[4] Einstein was perhaps the most famous German intellectual of his time. It is reasonable to assume that Einstein was familiar with Tagore's poetry and other writings. So, it is not surprising that the two might have sought each other out during Tagore's visit to Germany.

What is surprising is the topic of their conversation. We know that both Tagore and Einstein were deeply concerned with the rise of fascism and threat of war in Europe. Romain Rolland and others were actively engaged in organizing world-famous intellectuals against the war. It would have been a stellar topic of conversation for Tagore and Einstein. Instead, as recorded in the truncated published versions, they chose to discuss the character of science in the first meeting and music in the second.[5] For reasons of space, I will focus only on the first without denying that the second may have some bearing on the discussion of reality as well.

Even there, the Tagore-Einstein conversation does not really have the form of a debate. Rather, it has the form of repeated assertions by two minds reflecting in parallel. And it is not very clear what these assertions were about. For example, from some parts of the conversation it appears that the conversation was about whether there is a mind-independent reality. Dmitri Marianoff, who reported these conversations in the *New York Times*, titled the first dialogue, 'Thoughts on the possibility of [truth's] Existence without relation to Humanity.' It is suggested that Einstein says 'yes,' Tagore says 'no.' It is not clear that there was in fact such a direct opposition.

Consider the issue of the table. While Einstein clearly holds that the table continues to be there even if no one sees it, Tagore does not quite say that in that case the table won't be there. He says instead that it will be there but under the gaze of a universal mind. So Tagore does not deny the existence of the table when no individual human is present to perceive it; he ascribes the existence in that case to the presence of a universal mind. Similarly, in his *Three Dialogues between Hylas and Philonous*, George Berkeley held that since sensible things may exist independently of human beings, "there must be some other mind wherein they exist."[6] For Berkeley this 'other' mind is God; Tagore called it variously Supreme man, Universal being, and the like.

The standard criticisms against Berkeley thus apply to Tagore as well. The postulation of a constantly and universally aware universal mind does not seem to have more explanatory power than the simpler postulation of the table itself. The universal mind will need the table for a veridical perception of it in any case; if that perception is non-veridical, the table won't be there, which is a consequence both Berkeley and Tagore deny. So Tagore needs to postulate all of (a) universal mind in the form of all-pervasive consciousness, (b) ability of the human mind to grasp the universal mind *and* (c) veridical perception, just to say that the table continues to be there. Even though Einstein does not offer any further realist argument in favour of his view that the table continues to be there ("I cannot prove my conception is right, but that is my religion"), his robust and simple common sense seems to outweigh Tagore's complicated idealist thrust.

Philosophical difficulties aside, for the case under study, humans do seem to have a preference for the simpler option of settling for the reality of the table because that sense of reality is in fact withdrawn in some special cases. For one, even if, other things being equal, we robustly believe that the table continues to exist when I leave the room, we do not believe that my shadow – which also I can see among other 'thing'-like stuff – will continue to exist in the room if I leave the room. This is because other things are not equal; unlike the table, the continued existence of my shadow is tied to my presence in the room.

So, there is a clear distinction between things to which we do or do not ascribe reality even if they are perceptible.

To pursue the topic of shadows a bit more, it is interesting that humans not only make a thick distinction between independent objects like tables and mountains on the one hand, and 'dependent' objects such as shadows on the other, they make grades of distinction between the reality of shadows as well. For example, humans ascribe reality to shadows of individual persons even if the person is not present in the frame; they withhold such ascriptions if bottles are so arranged to generate a shadow-outline of a human face. These distinctions cannot be made sense of without a prior distinction between what is real and what appears to be so, *both* for the human mind.

Consider another – more classical – example requiring the appearance-reality distinction. When setting out in the dark in a snake-infested area, we take a torch along even if there is some moonlight. If I see what looks like a rope, I switch on the torch to double-check that it is in fact a rope, not a snake. Under those unfavourable conditions of light, we do not generally ascribe veridicality to the perception of rope without further examination. Since the 'universal mind' works throughout without getting affected by specific circumstances, a Berkeleyan has no further resources to explain these special cases. Einstein's realist conviction is thus ratified by robust common sense.

This is not to deny the appeal of Tagore's basic assertion that the truth which is described as scientific and which only can be reached through the process of logic – in other words, by an organ of thought which is human. However, Tagore's own explanation of what he means is not very helpful, as we saw. For Tagore, scientific truth obtains when the human mind forms a 'universal harmony' with the 'supreme being, *Brahman*.' Such an explanation seems unnecessary because the basic claim is obvious as scientific theories are unfailingly human products. Furthermore, the postulation of a universal being takes the explanation away from the *individual* human being whose perception of the table and the conception of scientific truth *are* the relevant phenomena. In that sense, Tagore in fact loses his grip on the 'humanness' of the issue.

3

It is not clear how this obvious point that scientific theories cannot fail to be products of the human mind touches on the classical issue of realism of scientific theories. We cannot take this point itself to lead to an anti-realist position because then the realist position cannot even be articulated. I don't think people such as Einstein, and more importantly philosophers of science like Karl Popper, who advocate the realist option even for (a completed) quantum mechanics, will want to argue for non-human origins of scientific theories. So the realism issue must arise, if at all, within the human enterprise.

I think a brief discussion of the realism issue is needed at this point basically to set it aside because there is a growing tendency, even among physicists, to view the 'debate' between Tagore and Einstein in terms of the realism debate around quantum theory. Tagore is said to be more inclined towards the uncertain, non-deterministic, subjectivist view projected by quantum theory in contrast to Einstein's classical determinist, universalist view of the physical universe. I hope to show that Tagore's concern was equally classical, different from the realism issue, and more fundamental.

Both the realist view of quantum mechanics via hidden variables, multiple worlds, implicate order, or whatever, and the anti-realist, instrumentalist Copenhagen view are to be understood within the undeniable idea that quantum mechanics is a human product, just as relativity theory and string theory are. The realism issue has to do with the character of particular scientific theories, not with the identity of the species that constructs those theories; there is just one species. As argued above, the issue has to do with whether we view the terms of a scientific theory to *refer* to mind-independent objects such as tables and trees or whether they are more like shadows.

For thinkers like Noam Chomsky,[7] the concept of reference in fact gets its salience from the over-riding scientific norm that scientific thinking aims to discover real joints of nature, notwithstanding the use of elaborate instrumental means such as mathematical models, approximations, artificial experimental setups, etc. This norm was the primary source of Einstein's stubborn realist conviction, as we saw, which was prepared to over-rule even as sophisticated a mathematical theory as quantum mechanics simply because it failed to indicate the required distinction between tables and shadows. This is because, on some popular Copenhagen-type interpretations, quantum phenomenon, such as diffraction of electrons through a double slit, is irreducibly explained in terms of a local observer. According to Einstein, the presence of the observer introduces the element of 'shadow' in quantum theory. Einstein concluded that such a theory must be 'incomplete.' For Einstein, the suggested humanness of all science does not thereby validate quantum theory. To emphasize, *humanness of science does not necessarily endorse subjectivism.*

The postulation of a mind-independent world is thus a human ideal – a 'norm,' pace Chomsky – which some scientific theories may fail to meet on occasion despite extensive use of the human cognitive apparatus. On that count, quantum theory is puzzling and problematic, according to Einstein and his followers,[8] precisely because there are scientific theories, such as Newtonian mechanics, falling under the so-called Galilean Style, that *do* meet the norm.[9] In effect, the human mind is such that, given a choice between general relativity and astrology, one is prone to ascribe truth and reality – and thus reference – to the former while denying them for the latter because astrology is contaminated with the (undesirable) 'human' factor: shadows. Again, this asymmetry in ascription of reality is something that the anti-realist finds hard to explain.

There is no contradiction between the essential human origin of scientific theories and the adoption of a realist norm. In fact, as Tagore insisted (as noted), a realist norm is not available unless scientific theories have a human origin: 'truth . . . must be essentially human; otherwise, whatever we individuals realize as true, never can be called truth.' We may safely conjecture that higher animals such as the great apes do form sophisticated views of the world in order to find strategies for survival. If these views (in fact) 'fit' the world, the apes survive, otherwise they perish. For the ape, there is no further reflective issue of *whether* the view fits the world. The ape, not unlike the postmodernist, need not have a concept of (objective) truth.

4

Earlier, I mentioned the value of Tagore's poetic conception of how things are such as the character of bird-songs. Naturally, then, Tagore's deeply reflective views on the nature of reality is likely to be represented more effectively in his literary work, rather than in occasional philosophical pronouncements of dubious value as we saw. Consider the first two lines of his very popular lyric – *mahavishye mahakashe* – set in *raga yaman kalyan* (I am not using the rest of the devotional part of the song; all translations from Bangla are by me):

> In the cosmos, the endless sky, the boundless time
> I, the human, travel alone in wonder.

The first line depicts what philosophers call the 'space-time manifold' consisting of all that is there in the universe located in the universal space-time framework. The second line 'humanizes' this manifold. I have access to the *entire* manifold for further reflection because I can be *there* at every possible point of this manifold as a part of my travel-itinerary. Tagore emphasizes the idea of *travelling in wonder*; he mentions it twice in succession. Perhaps he means that, if the manifold were to be my *construction*, an effect of my conscious imagination, then there is no travel, no wonder. I travel in wonder because the manifold is there independently of my reflection such that I *experience* the manifold just as I experience the table out there. Yet, this independence is meaningful just because I *grasp* it in wonder.

I think Tagore elaborates on the notion of a humanized cosmos at many places in his creative work. Consider another very popular lyric – *akashbhora surjyatara* – in *raga mishra kedara*.

1 The sky is full of the sun and stars, the world is full of life
2 I have found my abode there,
3 Thus my song comes alive in wonder.
4 The waves of eternal time that cause the ebb and tide
5 Also guide the rush of blood through my veins,
6 Thus my song comes alive in wonder.
7 When I walk on the grass on the forest-path,
8 The scent of the flowers startle my spirit
9 The gift of joy is scattered all around . . .

In line 1, Tagore begins to describe the cosmos as a sky-ful of suns and stars, but it also includes life in the universe. In line 2, just as in the earlier poem, he locates the I, the poet himself, amongst the elements of the cosmos. Line 3 says, this wonder of self-realized cosmos gives rise to the poet's song. In lines 4 and 5, Tagore highlights the unity of the cosmic order – in terms of gravitation – by noting that the reason why my blood flows is the same as why ebb follows tide for eternity. In lines 7 and 8, the idea of the cosmos is expanded to include blades of grass on the forest-path (7) and the smell of flowers (8), both giving rise to the joy and wonder (lines 6 and 9). Similar sentiments can be found in dozens of other poems and songs.

How do we interpret this image of a human universe? In some literary and philosophical circles, Tagore's colourful 'animated' conception of the cosmos is often over-interpreted to suggest parallels with the Upanishadik tradition, or the primacy of artistic (fictional) imagination. In both cases, the emphasis is placed on the subject, the I, as *constructing* the cosmos. In other words, the cosmos is viewed as *dependent* on the solitary self, the cosmos *disappears* if the self does. Tagore is thus viewed as a classical idealist. As noted, some of his inadequate philosophical arguments do suggest such a picture – wrongly in my view.

Tagore, as noted, did not develop his complex worldview with professional philosophical rigour. There is no doubt that his conception of the universe is a fallout of his artistic wonder; it is not a systematic product of cold analytical reflection. Tagore's conception of universe is scattered in a large body of artistic work whose meaning changed sharply throughout his life; in some cases, he did leave the impression of an idealist's universe. For example in one of his much-cited poems – *amar-i chetanar ronge* – he seems to assert that the emerald gets its colour due to my consciousness, the sky is lit up when I open my eyes, etc. Authors cite these lines with glee to emphasize Tagore's 'subjectivism' and affinity with quantum theory.

But it is routinely missed in the literature that, in this poem, Tagore carefully keeps to what philosophers call 'secondary qualities' after the seventeenth-century British philosopher John Locke – qualities of matter that seem to manifest only under human observation and are not intrinsic to matter. He also mentions aesthetic properties such as 'beautiful' – *golaper dike cheye bollam 'sundar'* – which of course have intrinsic reference to the beholder. However, there is no evidence that Tagore thought that the so-called 'primary qualities' such as size and density of objects – proper subject matters of physics – are also products of the beholder.

Be that as it may, let us assume that Tagore might not have been fully consistent in his worldview through his varied artistic expression. Yet, as explained in some detail above, a central part of his artistic work seems to project the conception of a human reality, which is metaphysically independent of the human mind, but gets its entire epistemic significance via human experience. This is the sense in which the universe is at once familiar and unfamiliar – *chiradibaser vishwa ankhisammukkhei*:

> The perennial universe in front of my eyes,
> I have seen a thousand times
> At my door.
> This timeless familiarity of the unfamiliar
> Has filled the deep recesses of my heart
> At ease.

So, the human universe is an *experienced* universe. Only in human experience does the universe carry its full significance since human experience is interpretive – 'stimulus-free,' to use a term from contemporary cognitive psychology. The universe appears to us in all its variegated complexity because humans – only humans – are endowed with 'organs of thought' that transform the experience into a symbolic conception. And an experienced universe has the self as the interpreter of experience at its cognitive centre. As Tagore puts it: the grand-designer works on the canvas of human ego – *manusher ahankar-pote*.

5

The idea that humans are experienced interpreters is an ancient one, although it was stressed with much force in the Western rationalist tradition. These philosophers pointed out that humans are not just passive receivers of external stimuli. The human mind actively contributes from its own inner resources to organize and interpret sensory information. The rationalist philosopher Ralph Cudworth called these resources 'cognoscitive powers', which enable the mind to raise 'intelligible ideas and conceptions of things from within itself.' The 'intelligible forms by which things are understood or known,' Cudworth held, 'are not stamps or impressions passively printed upon the soul from without, but ideas vitally protended or actively exerted from within itself.' For another rationalist philosopher Rene Descartes, the human ability to form conceptions of things from chaotic and often-impoverished experience is akin to 'a statue of Mercury contained in a rough block of wood.' The rationalists held that these cognoscitive powers, held together by human language, marked the basic distinction between humans and animals. The suggestion is that humans have a distinct (mental) design that generates the conception of a mind-independent reality.

Although the rationalists inaugurated the discussion, Immanuel Kant studied the design-factor with much ingenuity. He opens his masterwork *Critique of Pure Reason* (CPR) with the following suggestion:

> But though all our knowledge begins with experience, it does not follow that it all arises out of experience. For it may well be that even our empirical knowledge is made up of what we receive through impressions and of what our own faculty of knowledge (sensible impressions serving merely as the occasion) supplies from itself.[10]

Thus, Kant postulated a 'faculty of knowledge' – what Tagore called 'the organ of thought' – to account for human knowledge. As an aside, let me note that it was quite possible that Tagore had some familiarity with Kant's work since Kant was vigorously discussed in the intellectual circles of Bengal during Tagore's time. Specifically, Tagore's elder brother, the philosopher Dwijendranath Tagore, studied and wrote extensively on Kant and the relation of Kant's work with some aspects of classical Indian philosophy. There is some evidence of long philosophical discussions between the two gifted brothers.[11]

An immediate consequence of Kant's conception of human knowledge is that, since knowledge is necessarily linked to human receptive and interpretive faculties – *sensibilia* and *understanding*, respectively – the conception of a mind-independent reality, the noumenon, is logically incoherent since, by definition, we cannot describe it. In that sense, our conception of reality is restricted to what is intelligible to us, and intelligibility is directly related to our phenomenal grasp of the world. I return to the issue of intelligibility in the next section.

In the absence of a coherent conception of mind-independent reality and granting the existence of physics, how is physics possible? What does it mean to say that physics unearths the fundamental principles of the Universe if there is no Universe to inquire into in the first place? How do we explain the universal (*a priori*) empirical (*synthetic*) content of physics? How are *synthetic a priori* judgements possible? Notice that this is no longer a query *in* physics, but *about* physics. Tagore's emphasis on 'an organ of thought which is human' to appreciate 'the truth which is described as scientific' is, therefore, essentially a Kantian proposal.

Early on in his CPR, Kant poses the suggestion that a sense of mind-independent reality is a fundamental human endowment. There is no clear argument, but some hint can be found in his treatment of space-time, two of the fundamental concepts of physics. For Kant, both space and time are *transcendentally* ideal, that is, human contribution to the possibility of experience; space and time thus are products of *sensibilia* where space is the outer sense and time is the inner sense. In other words, space and time are viewed as real since that is how we are bound to think about the world. For example, he gives the following argument for the conception of *empty* space and time:

> For if we regard space and time as found in things themselves, and if we reflect on the absurdities in which we are then involved, in that two infinite things, which are not substances, must continue to exist, even although all existing things be removed – we cannot blame the good Berkeley for degrading bodies into mere illusion.[12]

So the only coherent solution is to 'assert then the *empirical reality of space* as regards all possible outer experience' while at the same time we 'assert its *transcendental ideality* – in other words, that it is nothing at all.'[13] Similarly, 'Time is therefore to be regarded as real, not intended as object but as the mode of representation of myself as object. That is, I really have the representation of time and of my determination in it.'[14] Space-Time are empirically real because without the required manifold we cannot make sense of the notion of experience without which the enquiry on how physics is possible cannot begin.

Similarly, the transcendental ideality of objects such as the table via the human mode of understanding must also project the empirical reality of the table because without this endowment we cannot make sense of the content of our experience of tables, as Einstein insisted. The required resources for doing physics thus arise from the human design itself within the constraints that any design imposes on the product. Tagore's 'Universal Mind' is part of the species-design of humans, and is thus located in each pathologically sound human mind.

6

The suggested humanness of science was also central to the issue of intelligibility and the conception of a Galilean science.[15] Galileo held that humans will never completely understand even 'a single effect in nature.' Chomsky suggests that Galileo's pessimism was validated by later developments in Newtonian science which 'not only effectively destroyed the entire materialist, physicalist conception of the universe, but also *the standards of intelligibility* that were based on it.'[16] Thus Chomsky supports Alexander Koyre's remark that 'we simply have to accept that the world is constituted of entities and processes that we cannot intuitively grasp.' The force of the expression 'intuitively' seems to be that we cannot have direct knowledge of how the world is like; the knowledge has to be routed in terms of resources available to our theory-building abilities – the Kantian view, as we saw.

Thus, any conception of the universe is restricted to what is intelligible to us; as standards of intelligibility fall, so does our grasp of the universe. The restriction gives rise to the old irony that the world, which undoubtedly gives rise to our knowledge of it, cannot be sufficiently grasped by the only means available to us. Chomsky suggests that, in a way then, we are compelled to adopt David Hume's position that Newton's discoveries reveal the 'obscurity' in which 'nature's ultimate secrets ever will remain.' The perspective seeks to question what is taken for granted, namely, that 'the natural sciences seek to discover basic truths about the world.' Clearly, these remarks apply to the whole of science including the most ingenuous proposals in theoretical physics. Notice that the scepticism arises due to the humanness of science.

If science cannot explain 'a single effect in nature,' how do we explain the sense of deep understanding, of genuine scientific explanation in some selective domains? How is it that some instances of science, say, theoretical physics, convey an abiding sense of truth, a view of 'the real properties of the natural world'?

The answer is located in the idea of mathematical physics. The mathematical formulae of physics are not only empirical in character, they also signal vast generalizations: Newtonian mechanics, relativity theory, quantum theory, and now string theory are often viewed as theories of everything. So, the really puzzling feature of the fundamental laws of physics is that they are at once mathematical in character and representations of large aspects of the universe.

This primacy of mathematical physics as the ultimate form of human knowledge was recognized since the beginning of modern physics. Johannes Kepler held that 'nature is always able to accomplish things through rather simple means, it doesn't act through difficult winding paths.'[17] Galileo Galilei thought that 'nature generally employs only the least elaborate, the simplest and easiest of means . . . nature is perfect and simple, and creates nothing in vain.'[18] Isaac Newton suggested that 'we are to admit no more causes to natural things than such as are both true and sufficient to explain their appearances . . . for nature is pleased with simplicity, and affects not the pomp of superfluous causes.'[19] Albert Einstein said that 'nature is the realization of the simplest conceivable mathematical ideas.'[20]

Authors such as Steven Weinberg[21] in fact trace the *realistic significance* of physics to its mathematical formulations: 'we have all been making abstract mathematical models of the universe to which at least the physicists give a higher degree of reality than they accord the ordinary world of sensations.'[22] Weinberg and others[23] have called this form of explanation in physics the 'Galilean Style.' The style, according to these authors, works as a foundational methodological principle in science, especially physics.

The discussion of intelligibility and mathematical physics brings out three salient aspects of a humanized conception of reality. These aspects may be ascribed to the emergence of Galilean style:

A Abstracting away from 'ordinary world of sensation,' 'most of what we find around us in the world of ordinary experiences is unhelpful for determining the real properties of the natural world';[24]

B Assuming the universe to have a 'simple,' 'perfect' design. Nothing else can be studied!

C Prioritizing mathematics, which is totally *a priori* and, hence, an exclusive product of the human mind.

7

When (empirical) physics is mathematized, the realist norm is recaptured within human intelligibility itself. In a very significant sense, then, the Galilean style essentially constructs an imaginary world. Yet, the amazing thing is that it actually works! The world does seem to obey simple mathematical principles. I will discuss two popular examples briefly.

The Fibonacci Sequence, discovered by the twelfth-century Italian mathematician Fibonacci, is the series of numbers: 0, 1, 1, 2, 3, 5, 8, 13, 21, 34, . . . The next number is found by adding up the two numbers before it: 2 is found by adding the two numbers before it (1 + 1); Similarly, 3 is found by adding the two numbers before it (1 + 2); and 5 is (2 + 3); and so on! The sequence is generated by a very simple *recursive* function that imposes the same order repeatedly at progressively higher levels: $x_n = x_{n-1} + x_{n-2}$. The series was also described by Indian mathematicians working on syllabic and metrical structure of speech hundreds of years before Fibonacci. Indologists suggest that 'the sequence Fn had already been discussed by Indian scholars, who had long been interested in rhythmic patterns . . . both Gopala (before 1135 AD) and Hemachandra (c.1150) mentioned the numbers 1, 2, 3, 5, 8,13, 21 explicitly.'[25]

The interesting aspect is that this progress of numbers maintains a constant ratio, the 'golden' ratio:

A B	B/A (Golden ratio)
2 3	1.5
3 5	1.6666 . . .
5 8	1.6
8 13	1.625
and so on.	

When the numbers are plotted figuratively, the golden ratio generates a systematic spiral. The amazing thing is that the Fibonacci spiral is observed in many forms in nature, most notably in the organization of seeds, leaves and petals in flowers and other flora. For example, it is observed in yellow chamomile head showing the arrangement in 21 (blue) and 13 (aqua) spirals.

The second, slightly more difficult, example comes from the study of fractals, self-replicating geometrical shapes. A simple and small geometrical shape such as a triangle or an open section of a curve may be repeated at different levels and scales to generate some of the most complex phenomena in nature. The 'self-replication' or 'self-similarity' can come in a variety of forms: exact self-similarity as in some snowflakes; quasi self-similarity noted usually in artificial mathematical figures such as Mandelbrot sets; statistical self-similarity as in shapes of coast-lines, and so on.

As with the Fibonacci sequence, fractals can be mathematically generated from simple recursive functions such as $z = z^2 + C$. A wide variety of fractals, both natural and artificial, are generated, repeating basic simple forms at a variety of levels and scales. From organization of leaves to snowflakes to coast-lines to solar systems, fractals exemplify Galileo's statement that 'nature generally employs only the least elaborate, the simplest and easiest of means . . . nature is perfect and simple, and creates nothing in vain.'

To emphasize, fractals and Fibonacci sequences, among many other similar recursive forms, are products of the ingenuous human (mathematical) mind and, as noted, mathematical ideas, by themselves, are entirely *a priori*, that is they are independent of the sensibilia. Yet, these beautiful 'fictions' seem to fill all parts of nature. It is in this specific sense of the mathematization of reality that we may understand Tagore's poetic idea that 'the human mind forms a universal harmony with the supreme being,' except that nature itself is that supreme being, no additional anthropomorphism in the form of *Brahman* is needed.

Needless to say, deep scientific inquiry into the order of nature, as in hundreds of years of mathematical physics, requires even more ingenuity, not only to describe nature, say, in terms of Fibonacci numbers, but to explain the inner structure of matter that gives rise to its surface systematic forms. That deeper inquiry, resulting in the postulation and discovery of fundamental laws governing basic elements and forces can only be 'reached through the process of logic which itself is an organ of thought which is human,' as Tagore emphasized in his poetic gesture. Humans, after exercise of ingenuity, come to grasp reality because reality, in that explicit sense, is human.

Notes

1 In Dmitri Marianoff, 'Einstein and Tagore Plumb the Truth: Scientist and Poet Exchange Thoughts on the Possibility of its Existence Without Relation to Humanity', *New York Times*, August 10, 1930.

2 Nirmalangshu Mukherji, 'Academic Philosophy in India', *Economic and Political Weekly*, 2002, 37(10) (March); Nirmalangshu Mukherji, 'Textuality and Common Life', in Supriya Chaudhuri (ed), *Literature and Philosophy*, Jadavpur University, 2005.
3 Nirmalangshu Mukherji, 2012. I sing my song (in Bangla) (*aami gaai gaan*). *Anustup*, Festival Number (*Sharodiya Sankhya*)
4 Martin Kämpchen, *Rabindranath Tagore and Germany: A Documentation*, Kolkata: Max Mueller Bhavan (Goethe-Institut), 1991.
5 Marianoff, 'Einstein and Tagore Plumb the Truth.'
6 Bishop George Berkeley, 1731. *Three Dialogues Between Hylas and Philonous, The Principles of Human Knowledge*, Collins/Fontana, 1962.
7 Noam Chomsky, *New Horizons in the Study on Language and Mind*, Cambridge: Cambridge University Press, 2000.
8 Karl R. Popper, *Quantum Theory and the Schism in Physics: From the Postscript to the Logic of Scientific Discovery*, London: Routledge, 1992.
9 Nirmalangshu Mukherji, 'Truth and Intelligibility', in Amiya Dev (ed), *Science, Literature, Aesthetics Volume XV, Part 3 of History of Science*, New Delhi: Philosophy and Culture in Indian Civilisation, Centre for Studies in Civilisations, 2009.
10 Immanuel Kant, *Critique of Pure Reason*, trans. N. K. Smith, London: Macmillan Company, 1787/1929, Paragraph B2.
11 Bijoy Mukherjee, '*Darshonik Dwijendranath* (in Bangla)', *Akademi Patrika*, 2007 (22), 9–48, Bangla Akademi, Kolkata.
12 Kant, *Critique of Pure Reason*, Paragraph B71.
13 Ibid., Paragraph A28.
14 Ibid., Paragraph A37.
15 Mukherji, 'Truth and Intelligibility.'
16 Chomsky, *New Horizons in the Study on Language and Mind*, emphasis added. Also, Wolfram Hinzen, *Mind Design and Minimal Syntax*, Oxford: Oxford University Press, 2006.
17 Johannes Kepler, *Astronomia nova* (New Astronomy), 1609.
18 Galileo Galilei, 1632. *Dialogue Concerning the Two Chief World Systems*, Berkeley: University of California Press, 1962.
19 Isaac Newton, *Philosophie Naturalis Principia Mathematica*, London, 1687.
20 Albert Einstein, *Ideas and Opinions*, New York: Bonanza Books, 1954.
21 Stephen Weinberg, 'The Forces of Nature', *Bulletin of the Americal Academy of Arts and Sciences*, 1976, 29: 13–29. Stephen Weinberg, *Dreams of a Final Theory*, New York: Vintage, 1993.
22 Weinberg, 'The Forces of Nature', 13–29.
23 Noam Chomsky, *Rules and Representations*, Oxford: Basil Blackwell, 1980.
24 Chomsky, *New Horizons in the Study on Language and Mind.*
25 Parmanand Singh, 'The So-called Fibonacci Numbers in Ancient and Medieval India', *Historia Mathematica*, 1985, 12(3): 229–44.

References

Berkeley, Bishop George. 1731. *Three Dialogues Between Hylas and Philonous, The Principles of Human Knowledge*, Fontana: Collins, 1962. Reprinted Penguin Classics, 1988.

Chomsky, Noam. 2000. *New Horizons in the Study on Language and Mind*, Cambridge: Cambridge University Press.
———. 1980. *Rules and Representations*, Oxford: Basil Blackwell.
Einstein, Albert. 1954. *Ideas and Opinions*, New York: Bonanza Books.
Galilei, Galileo 1632. *Dialogue Concerning the Two Chief World Systems*, Berkeley: University of California Press, 1962.
Hinzen, Wolfram. 2006. *Mind Design and Minimal Syntax*, Oxford: Oxford University Press.
Kampchen, Martin. 1991. *Rabindranath Tagore and Germany: A Documentation*, Kolkata: Max Mueller Bhavan (Goethe-Institut).
Kant, Immanuel. 1929. *Critique of Pure Reason*, trans. N. K. Smith, London: Macmillan Company.
Kepler, Johannes. 1609. *Astronomia nova* (New Astronomy). Reprinted Santa Fe: Green Lion Press, 2015.
Marianoff, Dmitri. 1930. Einstein and Tagore Plumb the truth: Scientist and Poet Exchange Thoughts on the Possibility of its Existence Without Relation to Humanity. *NewYork Times*, August 10.
Mukherjee, Bijoy. 2007. Darshonik Dwijendranath (in Bangla). *Akademi Patrika* (22): 9–48, Bangla Akademi, Kolkata.
Mukherji, Nirmalangshu. 2012. I sing my song (in Bangla) (*aami gaai gaan*). *Anustup*, Festival Number (*Sharodiya Sankhya*).
———. 2009. Truth and Intelligibility. In Amiya Dev (ed), *Science, Literature, Aesthetics Volume XV, Part 3 of History of Science*, New Delhi: Philosophy and Culture in Indian Civilisation, Centre for Studies in Civilisations.
———. 2005. Textuality and common life. In Supriya Chaudhuri (ed), *Literature and Philosophy*, Jadavpur University.
———. 2002. Academic Philosophy in India. *Economic and Political Weekly*, 37(10), March.
Newton, Isaac. 1687. *Philosophie Naturalis Principia Mathematica*, London. Reprinted Los Angeles: University of California Press, 2007.
Popper, Karl R. 1992. *Quantum Theory and the Schism in Physics: From the Postscript to the Logic of Scientific Discovery*, London: Routledge.
Singh, Parmanand. 1985, The So-called Fibonacci Numbers in Ancient and Medieval India. *Historia Mathematica*, 12(3): 229–44.
Weinberg, Stephen. 1976. The Forces of Nature. *Bulletin of the American Academy of Arts and Sciences*, 29: 13–29.
———. 1993. *Dreams of a Final Theory*, New York: Vintage.

2

TOO BIG A BANG FOR LANGUAGE

Tagore's critique reloaded

Probal Dasgupta

Methodological reasons make it appropriate to open this chapter with a self-conscious non-citation. I would have liked to begin by citing something the American playwright Arthur Miller said in early 1977 or possibly in late 1976. But relevant documentation is emphatically unavailable. An in-flight magazine provided by the Icelandic airliner on my flight from New York to Reykjavik in summer 1977 featured an interview with Miller, who had recently been to China; I am unable to recall how recent the trip had been relative to the publication of the interview, so I'm hedging my bets on the date of Miller's China visit. *What was it like*, they asked Miller. *It was impossible to get anything across*, he replied; *there was too much cultural shouting*.

I am citing this remark from memory, but I'm quite sure about the key phrase *cultural shouting*. Now, unless some quixotic hunter catches and delivers the elusive chapter and verse citation, you will have to treat my report of Miller's words as hearsay, a mode of citation inadmissible in rigorous discourse. Given the theme of this volume, and given the ubiquitous belief that science is rigour writ large, it should be clear why I choose to open the chapter with an allusion to cultural shouting – an allusion whose inadmissibility as a rigorous reference I underscore explicitly. To fortify this methodological point, I hereby request you not to bother your friends in Reykjavik. Even tracing that interview in that magazine will not convert what I've just said into an academically acceptable reference. What self-respecting scholar would ask colleagues to treat in-flight reading as serious source material?

Throwing methodological caution to the winds – it is the *winds* tucked away in this turn of phrase that will hold our attention throughout this chapter – let us now take a closer look at this notion of cultural shouting that Miller invokes in his remark. I take it that he is using this unusual locution to mean that the people he tried to have a conversation with kept saying, '*We in China* think this way; this is *our culture*;

what is *your culture*? What do *you people in America* think about this, or that, or the other?' Their zealously expressed belief (I construe 'shouting' as his term for such zeal) that all individuals are locked into their respective cultures frustrated Miller's attempts to speak softly – to initiate meaningful communication rather than a mere exchange of ritualized, repetitive, predictable formulas.

Please forgive me for saying this with a studied slowness, appearing to assume that my reader is not quick on the uptake. When one is trying to exchange thoughts across steep barriers, it pays to proceed slowly. For there is no good reason to assume that one must be making sense to those on the other side of the relevant barrier. The risks of slowness, easily noticed and guarded against, are less extreme than the perils of 'cultural shouting.'

In this chapter, I would like to make two points about the Tagore-Einstein exchanges, without engaging with the secondary literature to which many authors have made significant contributions, and without prejudice to the importance of what they have said. My first point is broad; it may serve to bring out the relevance of 'cultural shouting,' and to offer a cue for continuing the conversation beyond what is offered here. The somewhat specialized second point is presented in the form of an extended argument. There too I try to couch the issues in a way that should make it easy to take the discipline-specific debate further. Since I cannot afford to be slow throughout, I cut some corners to make my points without the extra delay that technical fortification would cause; therefore, the few linguistics-proficient readers who will engage with this chapter are invited (in the footnotes) to consult other writings of mine for a fuller unpacking of arguments. The nature of our context here leads us to omit the elaborations or defence that would have been obligatory in a disciplinary setting.

Point one: look ma, no force

Consider some Tagoreans and some Einsteinians today who wish to pick up the thread of the old exchanges on the basis of relevant material (material available in principle at that time that did not in fact shape the conversations between Einstein and Tagore). Team T and Team E, as we shall call them, would like to launch a fresh conversation that can lead in principle to significant agreement on issues that were crucial in Tagore's work as well as Einstein's – steering clear of areas of known disagreement.

Someone who has the time and inclination to draft that imaginary conversation on Team T and Team E's behalf may find some cues in

Russell's remarks on what he calls the *abolition of force* in Einstein's theory of gravitation. His comments appear in a book, which, though published long after 1930, rests on older, pre-1919 material. Russell writes (boldface mine): 'The supposed necessity of attributing gravitation to a "force" attracting the planets towards the sun has arisen from the determination to preserve Euclidean geometry at all costs. If we suppose that our space is Euclidean, when in fact it is not, we shall have to call in physics to rectify the errors of our geometry. We shall find bodies not moving in what we insist upon regarding as straight lines, and we shall demand a cause for this behaviour. . . . *If people were to learn to conceive the world in the new way, without the old notion of "force," it would alter not only their physical imagination, but probably also their morals and politics*. The latter effect would be quite illogical, but is none the less probable on that account. In Newton's theory of the solar system, the sun seems like a monarch whose behests the planets have to obey. In Einstein's world there is more individualism and less government than in Newton's.'[1]

Readers already clued into the terms of discourse in Einstein's and Tagore's approaches will immediately notice how close these ideas are to the latter's mid-life book of poems *Naibedya*,[2] in which he self-consciously, and with his father Debendranath Tagore's explicit and enthusiastic approval, refashions certain crucial theological settings on which Debendranath's version of the Brahmo revision of orthodox Hindu doctrine had been based. In this unjustly neglected book, Rabindranath Tagore eliminates the old construal of awe as literally **fear** of God. He rearticulates hierarchical respect – which under the *ancien régime* remains a sentiment designed to legitimate authoritarian systems as social reflections of the intrinsic superiority of the divine and of various rulers (hierarchical respect draws its validity from the claim that their right to rule rests on divine sanction) – by postulating in its place an obligatory mutuality of respect among human beings. Tagore claims that such a configuration alone, conferring universal rights on every person regardless of the boundaries of the empires ruled by white men (this counted as an advanced idea in 1901), would adequately reflect the injunction to reserve human awe for the divine and only for the divine.[3] Tagore's condemnation of every politics based on force was coupled with his promotion of a new culture (he took the position that it was best articulated as a set of locally driven non-political movements, given the poisoning of politics by the might-is-right propaganda emanating from all the power centres and their elite spokesmen) that would rest on non-authoritarian persuasion appealing to universal moral sentiments. These views consistently inform

Tagore's words and actions and are clearly akin to the moral and political associates that Russell's comment sees as natural companions for Einstein's take on gravitation – a take that depicts gravitation as a matter of the geometry of space-time and not as an application of force.

I leave to others the exercise of translating these suggestions into an imaginary conversation that takes the Tagore-Einstein dialogue beyond the cul de sac we have all been examining for nearly a century now. What we can now do in this text is find a way to implement seriously the Tagorean principle of mutual respect – and to do so at the boundary that continues to divide inquiry into 'the sciences' and 'the arts.'

Preamble to point two: extending the debate into scientific analysis

We know about Tagore's conversation with Einstein in which he conceptually articulates the idea that at the level of 'universal man' (a universal subject of consciousness whose view, as finite persons imagine it, provides the referential frame for all discourse) there is no basis for holding some purely truth-handling cognitive faculty at a distance from some affective supplement that the imagination of finite humans in its avoidable subjectivity attributes to that referential frame and associates with beauty. However, readers who have access only to Anglophone (or translated but Anglophone-derived) sources cannot possibly be aware of the fact that Tagore did not provide this conceptual articulation from the standpoint of an exclusively artistic craftsman of words and images. He was also a scientific analyst, whose work provides an empirical implementation of his perspective and is open for scrutiny by colleagues in his time and in ours. The domain he contributed to was linguistics.

Some readers are bound to contest the claim that this is relevant. There exists a widespread belief that the study of language does not belong under the rubric of scientific inquiry. Among those who subscribe to this belief, a significant section maintains that only one relatively recent approach to the study of language – known as generative grammar and comprising not just the work of Chomsky and his fellow thinkers but also several schools of thought that differ from his views only in certain technical details – counts as truly scientific, while all other approaches to the study of language fall outside science. Under such assumptions, Tagore's writings about language obviously become irrelevant. For he died in 1941, well before the advent of generative grammar.

In order to provide a clear line of argumentation that will be easy for readers to follow and thereafter to assent to or dissent from, I am making the following assumptions, (a)-(d), in order to stay as close to 'mainstream' linguistics as possible. My purpose in doing so is to eschew irrelevant controversy that would merely have obscured the core issues. Readers interested in writings that defend or contest (a)-(d) will have to look elsewhere. They will find that these views are accepted by most practising linguists today. They will also find that practitioners of the established natural sciences who sometimes informally express scepticism about the scientificity of linguistics (and who are likely therefore to reject point (d) below) have yet to translate their a priori scepticism into a rigorously argued view on the basis of an engagement with the results of contemporary formal linguistics. My stand is that their sceptical remarks merely express unexamined prejudices and must be set aside if serious discussion is to take place. Here, then, are the assumptions I am making:

a Tagore wrote in what linguists of our times call a 'theory-neutral' perspective. He did not even fit in the canons of any school of diachronic linguistics that flourished in his own period of history.
b Some of Tagore's observations broached what hindsight allows us to portray as issues dear to the generative grammarian's heart. The work of mining his texts to find and develop these observations[4] is far from over.
c The history-focused or diachronic approach to linguistics prevalent in Tagore's times led to structural linguistics and later to generative grammar through a 'natural' methodological succession whose legitimacy claims we choose not to question here. Generative grammar, despite contestations, is now the dominant approach to the systematic study of language. All earlier writings, even if embedded in the research programmes of historical linguistics or structuralism, become precursor material that can at best be recycled for data that generative grammarians can characterize in modern terms. Tagore's work, which did not even follow those canons rigorously, counts *a fortiori* as precursor material of this sort.
d Generative grammar emulates the methodological style of contemporary research in the hard sciences, especially physics. The details of just how linguistics is to be connected to other natural sciences are contested. But the broad consensus, which shapes the mind-set of its practitioners, has converged on the idea that generative grammar validly characterizes linguistic *ability* as

> a species-specific and species-uniform human trait – a research thread that belongs to the natural science side of the science-arts divide – while the study of language *use* falls under social, literary and other modes of inquiry.

Taking (a)-(d) as my point of departure, I shall, in the body of this chapter, argue that even generative grammar, when one follows its own logic all the way, leads to conclusions that undermine a **naive Galilean** approach to science (consonant with what Einstein was saying in the conversations we are focusing on) and strengthen a **sensitive Galilean** approach closer to Tagore's.

Point Two A

I do not propose to take up here the exegetic task of pinpointing specific descriptive proposals in Tagore's writings; situating them with respect to independently defensible linguistic frameworks; reformulating Tagore's proposals in terms of the apparatus of the frameworks with formal rigour; summarizing the result of these operations in a style that general readers will find user-friendly; and not overstepping the boundaries of the average attention span. Such a challenge will keep some of us occupied for the next 60 years or so, if scholars with appropriate qualifications decide to take it on. The goal of this chapter is to persuade some readers to kickstart that 60-year process. It is unrealistic to expect anybody reading the chapter today to do so; the point of leaving something like this in writing is to reach future readers willing and able to face the challenge.

Under the following subheadings 2Ai, 2Aii etc., I (i) exhibit a sample of Tagore's linguistic writing, (ii) outline my construal of his point, (iii) show how his point is relevant to our project in this article, (iv) explain why I choose not to rearticulate Tagore's own arguments here and (v) prepare you for 'Point Two B.' The subheadings should help you to follow this itinerary.

Point Two A (i)

I begin this subsection by exhibiting a relevant sample of Tagore's linguistic analysis, drawn from his 1938 book *Bāṅlābhāṣā-paricay* ['An introduction to the Bengali language']:[5]

> In Sanskrit we have derivatives featuring the suffix *ta*, as in *vikasita puṣpa* 'a flower that has blossomed', conversational

Bangla equivalent *phoTa phul*. In Bangla examples like *buk-phaTa kanna, cul-cera tɔrko, mon-matano gan, nuYe-pɔRa Dal, kuli-khaTano bæbSa*[6] 'heart-breaking weeping, hair-splitting argument, spirit-uplifting song, bowed-down branch, coolie-exploiting business', one observes the affixes **a** and **ano** [*set in **bold type** here to maximize inspectability*]. They are adequate, but do not handle complexity particularly well. For *a.cintita.pūrva ghaṭanā* 'an event not anticipated, lit. not. thought.before',[7] authentic Bangla has no concise equivalent.

'It must also be pointed out that conversational Bangla has several tournures that are unique to it. Colloquial Bangla often doubles a word to convey a nuance that no word with articulate or referential meaning can carry. It is impossible for articulately meaningful words to replicate the content, for instance, of *maTh **dhudhu** korche, roudro korche jhãjhã* 'the meadow is like *dhu* [a wide open empty space],[8] the sun-light is like *jhã* [very bright].' Expressive sounds are easier for the mind to grasp than articulated meanings, after all. Expressions like *uSkhuS* ['fidgeting' – remember that note 8 is applicable throughout], *niSpiS* ['itching to do something'], *phælphæl* ['staring'], *kãcumacu* ['abashed'] have no specifi-able meanings. It is as if one is extracting meaning from them as surplus value devoid of any grammatical imprimatur.

'Bangla has another sort of word doubling which carries a semblance of meaning, but gives primacy to suggestion rather than reference. Where Sanskrit [**and high Bangla** – a sup-plement that is implicit throughout this passage every time Tagore mentions Sanskrit by name: in this book he introduces the term *saṃskṛta Bāṅlā* for 'high Bangla, studded with San-skrit words'] gives us *patanonmukha* 'on the verge of falling,' Bangla [by which in this context Tagore means 'conversa-tional Bangla'] says *pɔRo-pɔRo* 'lit. fall.*o*-fall.*o*' [no definite grammatical or semantic value attaches to this *o*, for which therefore one cannot even offer an approximate gloss]. What is *āsanna* 'imminent' is Sanskrit is *hɔbo-hɔbo* 'lit. will.happen-will.happen' in Bangla. Likewise: *gælo-gælo* 'lit. gone-gone [he/she/it is on the verge of disaster, all hell is breaking loose],' *jaY-jaY* 'lit. goes-goes [he/she is about to die].' Sanskrit *vāṣp. ākula* '[eyes] misty [with tears about to be shed]' comes out as *kãdo-kãdo* 'lit. weep.*o*-weep.*o* [about to cry]' in Bangla. For Sanskrit *avaruddha.svar.e* 'choked.voice.Loc [in a voice choked with emotion]' the Bangla equivalent is *badho-badho* *gɔla.Y* 'stop.*o*-stop.*o* voice.Loc.' Such wording in Bangla conveys not just an idea, but a vivid image. Let's toss some verses together:

ja.bo-jabo kɔre, cɔron na Sɔre,
['go.will-go.will does, feet Neg move,']
phir.e-phir.e caY piche,
['turn.ing-turn.ing looks behind,']
pɔR.o-pɔR.o jɔl.e bhɔr.o-bhɔr.o cokh
['fall.o-fall.o tears.Loc fill.o-fill.o eyes']
Sudhu ce.Ye thake nice.
['steadily gaz.ing stays downwards.']

[approximate free translation:
She's about to leave, but her feet won't move,
She keeps turning around,
Her eyes, brimming with tears she's about to shed,
Steadily gaze at the ground.]

'If what an author wants to produce is a vignette in the line drawing mode, there is no choice but to use this hesitant diction [what is rendered here as 'hesitant' is Tagore's word *badh.o-badh.o* 'hesitate.*o*-hesitate.*o*']. My book *Śabdatattva*[9] ['The study of words'], in its discussion of sound symbolism,[10] elaborates on this use of expressive devices – 'shades of expression,' so to speak, that bring out clearly the 'shades of meaning' in graphic descriptions.

'Some Bangla affixes have breached the boundaries of referential meaning and moved into this shaded region. Consider the affix we find in the following set of examples: *kipTemo* ['stinginess,' from *kipTe* 'stingy'], *chiblemo* ['flippancy,' from *chible* 'flippant,' an archaic variant of the word *chæbla* that is still in use today], *chelemo* ['immaturity,' from *chele* 'boy'], *jæThamo* ['precocity plus uppitiness,' from *jæTha* 'elder paternal uncle'], *Thæ̃Tamo* ['insolence,' from *Thæ̃Ta* 'insolent'], *phajlemo* ['vapid frivolity,' from *phajil* 'vapidly frivolous'], *biTlemo* ['mischievousness,' from *biTle* 'mischievous'], *pejomo* ['wickedness,' from *paji* 'wicked'], *hænglamo* ['needy gluttonousness,' from *hængla* 'needy and gluttonous'], *bokamo* ['stupidity,' from *boka* 'stupid'], *bãdramo* ['monkeying,' from *bãdor* 'monkey'], *gõRamo* ['rigid orthodoxy,'

> from *gõRa* 'rigidly orthodox'], *matlamo* ['drunkenness,' from *matal* 'drunk'], *gunDamo* ['thuggery,' from *gunDa* 'thug'].
>
> 'What Sanskrit suffix can I possibly compare this suffix with? We can use *-tva* [pronounced *-tto* in Bangla] and consider glossing *kipTemo* 'stinginess' as *kipTetto*. But there is no affect attached to the [Sanskrit] suffix *-tva*; it is insensitive to distinctions between good and bad, preferred and dispreferred, animate and inanimate. And yet even a quick look at the list given above will convince you that these words are completely unacceptable in well-bred company. It is as if one had collected the choicest muck to be able to hurl insults at one's targets. Our affix – in its two alternants *mo* and *amo* – allows us to derive *bãdramo* ['monkeying,' from *bãdor* 'monkey'], but not to derive *Singhomo* ['lioning,' from *Singho* 'lion']. While *kipTemo* ['stinginess,' see above] is a well-formed derivative, *datamo* ['generosity,' from *data* 'generous'] is not. One can easily say *pejomo* ['wickedness,' see above], but one is unwilling to say *Sedhomo* ['saintliness,' from *Sadhu* 'saint'] for [the well-formed and current word] *Sadhutto*. I doubt that any other language reserves an affix exclusively for the function of venting one's wrath.
>
> 'Now look at another suffix, *pɔna*: *buRopɔna* ['oldmannishness,' from *buRo* 'old man'], *nækapOna* ['affectedness,' from *næka* 'affected'], *chiblepOna* ['flippancy,' from *chible* 'flippant,' an archaic variant of the word *chæbla* that is still in use today], *adurepɔna* ['indulgence-craving, brattishness,' from *adure* 'spoiled'], *ginnipɔna* ['housewifery,' from *ginni* 'housewife']. They are all about casting aspersions. One would have expected grammars to keep their affixes neutral, impervious to affect boundaries; but this suffix shows no interest in satisfying such an expectation. These devices look as if they have been sharpened for the express purpose of sitting around with one's friends in one's favourite haunt and eloquently speaking ill of the other side.'

After these remarks, Tagore proceeds to consider further affixes that exhibit the same properties. Readers familiar with Bangla can access the text themselves; as for readers unfamiliar with the language, experience shows us that they would not find it helpful if we were to prolong this excerpt. These considerations make it appropriate to end the subsection here, and to resume our discussion.

Point Two A (ii)

It is possible that the excerpt from Tagore's book may strike some readers as being concerned with an only locally relevant contrast between 'Sanskrit' (or the Sanskrit-laden erudite, high register of Bangla) and conversational Bangla. But his book was intended, in fact, as a critique of grammar itself as a comprehensive enterprise. Linguists in general, well into the 1930s, had consistently taken the position that grammar was an optimal toolkit capable of dealing exhaustively with all components of the task of describing a language; mutatis mutandis, even mainstream linguists in the twenty-first century tend to take this stand. But Tagore contests this position; he points to significant and pervasive phenomena that elude the grammatical toolkit even in its incarnation as the apparatus of *comparative and historical grammar*, which enjoyed scientific prestige well into the twentieth century; he makes concrete proposals to address these issues. As you will have noticed in the fine print of the excerpt, it is not just in 1938 that Tagore highlights these issues – some of his earliest linguistic writings dealt with them, as far back as 1900 (see note 10).

In order to see why Tagore felt he was entitled to regard grammar in general and the derivation of words from Sanskrit in particular as emblematic of systematic scientific analysis, we need to note that he dedicates his 1938 book to the linguist whom he took most seriously (and to whom in the text of that dedication he gives the honorific title *bhāṣācārya*, for which 'Sprachenlehrer' is perhaps a viable rendering) – Suniti Kumar Chatterji (1890–1977), best known as the author of *The Origin and Development of the Bengali Language*.[11] That monumental work offered a neogrammarian account of the history of Bangla. ODBL, as the book is usually called, presented Bangla as a language whose sound system, vocabulary and word formation devices were inherited from Old Indo-Aryan (for readers not clued into technical linguistics, it is only a mild oversimplification to equate this term with 'Sanskrit,' broadly construed). Admixtures from other sources were acknowledged, to be sure, but the principal line of derivation was from Old Indo-Aryan through Middle Indo-Aryan. Throughout his writings, Chatterji, like other linguists writing about Bangla and its sister languages at that time, took it for granted that every language could be comprehensively characterized in terms of the grammatical, lexicographic and etymological methods that classical and modern descriptions of Sanskrit had always depended on. His work was firmly anchored in grammar in general and comparative grammar in particular.

We cannot afford to read Tagore's book as just a response to Chatterji's work written by a gifted amateur admiring (but, in certain respects, gently questioning certain details in) the views of the stellar professional linguist of his time and place. There was a specific interpersonal context that needs to be brought out explicitly here.

Tagore, who in his youth was curious about matters of language and read everything that he could find on comparative grammar and the history of Bangla, naturally made a point of following Chatterji's progress when the latter came to the fore as the rising star on the firmament of historical linguistics in India. Immediately after the publication of ODBL, Tagore chose Chatterji as his travelling companion for an extended trip to Southeast Asia in 1927. Both Tagore[12] and Chatterji[13] wrote books recounting their experiences in Indonesia and some of their conversations about cultural, linguistic and religious history. Tagore's 1938 book, when we view it as a response to Chatterji's views – which for him represented the state of the art in the systematic study of language – has to be set in the context of their friendship and immense mutual respect at the personal level. These considerations help us to see that Tagore found it necessary to express his fundamental dissent about the whole approach gently, to avoid hurting a personal friend. My point is that his critique would probably have been expressed in sharper terms if friendship had not stayed his hand.

Point Two A (iii)

I hope to have shown that what looks like a minor local disagreement about how to address the linguistic phenomena of Bangla must actually be read as a fundamental critique by Tagore of the entire approach to historical linguistics that Chatterji embodied. However, I have yet to convince you:

(x) that Tagore wanted his 1938 book to be read as an extension of his take on the scientific investigation of the physical universe (an approach that he regarded as sharply distinct from Einstein's approach), and that therefore the material presented here is indeed relevant in the context of making sense of the Tagore-Einstein disagreements;

(y) that you are wrong to feel certain that, even if it turns out that Tagore did make a valid and relevant point or two, nonetheless 'mainstream' linguists today have nothing to worry about (since the controversy touches only some troublesome peripheral phenomena, and mainstream linguists know how to stay out of trouble), and that a fortiori non-linguists who are scientists or have faith in Einsteinian science need not take the present article seriously; and

(z) that the widespread belief (which you quite probably share) that a methodological abyss separates historical linguistics from the natural sciences (an abyss that renders Tagore's critique of Chatterji obsolete because linguistics has long moved away from the spot he is targeting), is ill-founded.

I address the first of these matters, (x), immediately; the second, (y), in the section 'Point Two B'; and the third, (z), in my 'Concluding methodological remarks.'

We know from Tagore's explicit prefatory comments, addressed to 'student readers,'[14] that he regards this 1938 book – a readable introduction to the world of human language based on his own approximate understanding of the state of the art in linguistic science – as a language-focused sequel to his 1937 book *Viśvaparicay* 'introducing the cosmos,'[15] which too belonged to the 'popularization of science' genre. There is no text other than these two books that Tagore ever identified as forming part of that enterprise. Other readers may wish to contextualize his remarks differently. But the fact remains that these remarks need to be taken into account when we reread the material today.

Point Two A (iv)

Why do I not choose, in this chapter, to rearticulate Tagore's own arguments, place the material in today's discourse, and relate the discussion to the Tagore-Einstein disagreement? Let me give a couple of examples of what would have been involved. I hope that these examples will help you to understand why I shall – under Point Two B below – move the discussion into territory not covered by Tagore himself, but making his point at a level that speaks to concerns now current in official linguistics.

Since this chapter is intended for a general audience, I chose a particularly accessible sample of the text at Two A (i), and can correspondingly provide a user-friendly rearticulation in terms of English words familiar to all of you.

I begin by noting one point emphasized by linguists today that was not quite as clear in Tagore's time. Namely, a particular affix may appear in a 'frozen' form in certain words but as a 'living' force elsewhere; in such cases, it is the 'living' affix whose trajectory we need to follow, letting sleeping (or frozen) dogs lie. To see this, consider an English suffix whose vitality nobody can possibly question – the suffix *ness*. In words like *madness*, *sadness, stillness*, *loudness*, this element routinely forms abstract nouns of quality from an adjectival base. However, in the words *wilderness* and *fastness*, we find a 'frozen' version of *ness* – it flags not a quality, but a type of place. How do we diagnose the status of the affix in these words – as a 'frozen' affix? We do this on the basis of the fact that we cannot freely attach this affix, carrying this sense, to other adjectives and form place-flagging nouns. We find ourselves unable to call a rocky place a 'rockiness,' or a marsh a 'wetness.'

Keeping this point in mind, let us now take up some English examples that illustrate phenomena more or less akin to Tagore's observations for Bangla. Consider words like *quackery*, *crookery*, *buffoonery*, *clownery*, *goofery*, *trickery*, *fakery*, *robbery*, *knavery*, *mockery*, *bribery*, featuring the suffix *ery*. They all have a negative affect to them, don't they? Now, one might try to counter this initial impression by pointing to affect-neutral examples like *confectionery*, *perfumery*, *shrubbery* or *greenery*. However, it seems clear that those instances of *ery* are 'frozen.' I would wager that if biotechnology were to produce a new range of vegetables that are distinctively blue, they wouldn't get called *bluery*; and that even a risqué shop marketing fumes as some sort of enhancer of beauty or health would think twice before launching the term *fumery* even at the level of a brand name. The pattern that counts as 'living' and extendable is the one associated with the negative connotation. Lexicographers have found people saying *apery*, *freakery*, *foxery*, *assholery*. If you go through a typical list of new words ending in *ery* (I bow to your superior web surfing skills as you hunt for such specimens, needless to say), you will find many vivid examples of people using the affix to vent their negative feelings. I don't suppose you will find people taking a base associated with a positive affect – a typical word like *saint* or *champ* or an untypical one like *mensh* or *woke* – and coining *ery*-words that preserve this positive tone. I'll be surprised if any readers of this chapter can imagine saying *saintery*, *champery*, *menshery*, *wokery* with a straight face to express adulation.

From *ery* we turn to *ish*. This affix is not straightforwardly negative. Obviously our first step is to set aside the frozen instances that appear in *Swedish*, *Danish*, *a puckish sense of humour*. However, as a living affix, it fans out into two different types of use. When applied to an adjective like *black* or a numeral like *fifty* it yields *blackish*, *fiftyish*, which are emphatically **vague**, and cannot be accurately called negative. However, when it is attached to a noun base, *ish* happily chooses a negative partner noun or adds a negative effect of its own, as in *boorish*, *selfish*, *bookish*, *impish*, *clownish*, *clannish*. And it firmly avoids positive company. Even if (within the relaxed register where lexical innovations find the most hospitable setting) we stay away from terms of adulation and consider only half-way positive words like *geek*, *nerd*, *champ*, we burst out laughing when we even imagine saying *geekish*, *nerdish*, *champish*. Our verdict must be something like 'on the whole, negative.'

Compare these low register compatible affixes with the slightly artsy, slightly pompous suffix *esque*, as in *Altmanesque*, *Hitchcockesque*, *Gandalfesque*. When we consider non-proper bases, we get words like *accordionesque*, *animalesque*, *apartheidesque*, *balladesque*, *billboard-esque*, *bimboesque*, *carnivalesque*, *cartoonesque*, *centauresque*. It's a mixed crowd, a lot closer to what Tagore said about Sanskrit affixes, approaching the ideal of an affix that is just a grammatical tool and doesn't care what affective connotations are associated with the base it attaches to. Likewise, the upper register affix *ity* appears both in positive *sagacity* and in negative *mendacity*: it never appears to favour bases associated with any particular connotation or affect.

In the region of qualitative affixes, then, English seems, judging from this tiny sample of two low register suffixes vs two high register affixes, to echo the overall configuration that Tagore finds in Bangla. Namely, the more colloquial affixes, the ones associated closely with words used intimately by ordinary people throughout the speech community, are distinctly more negative in terms of affective connotation than high register affixes, which adhere more closely to the prim and proper behaviour one expects of a cold, affect-free tool of the grammatical apparatus.

If you go back to the Tagore excerpt, you will find that this was only the last point he made; that he had a great deal to say about word doubling (hardly a feature well represented in the core grammar of English, French, German, Russian, Spanish); that he placed all his points under the rubric of suggestive and connotative associates of canonical lexical meaning; that the web of observations he was weaving for Bangla and seeking to anchor in his take on human language in general cannot be straightforwardly recast, with data from widely familiar languages in mind, in terms of any approach that either a non-technical or a linguistics-trained audience today can be expected to understand without an enormous investment of specific effort – for which the energies are obviously unavailable. The point of showing you a tiny sample of English data mirroring one of his observations is only to indicate very roughly what kind of spadework would be called for if I were to try to get his characterization of Bangla, or of human language, to even begin to make sense to a readership such as the one addressed in this chapter. And, alas, the audiences available for technical writings in linguistics are not in this respect better equipped than the heterogeneous and therefore non-specialist readership of the present chapter.

If someone were to make the Herculean effort and produce a heavily annotated translation of Tagore's linguistic writings with an Anglophone audience in mind – specialist or non-specialist, it hardly matters – the results would be predictable. The general public would wait for trained Anglophone linguists to review the book. The latter would read the first few pages and exclaim,

> This Tagore person provided an array of examples of onomatopoeia, of expressive lexemes, of reduplication and other expressively charged syntactic constructions, of discourse particles, and of other material that slips through the cracks of the conventional lexical and grammatical architecture of synchronic and historical linguistics. He seemed to imagine that the phenomena that interested him, if he showcased them properly, could somehow add up to a call for an alternative framework that would be sensitive to the phenomena close to his heart. We are underwhelmed. Contemporary linguistics does have room for alternatives to the formalistic mainstream. The cognitive approach to metaphor and other intermodular phenomena, the study of iconicity, the systematic investigation of expressives and reduplication – all this and more is available. Finding such regional articulations today that can contextualize Tagore's material in a comparative context, category by category, is a simple bibliographic exercise if somebody wishes to undertake it. That exercise may serve some cultural purpose, such as motivating speakers of Bangla to join the linguistics brigade. But there is no scientific use for this material.

Reading dismissive reviews along these lines by professional horses from the normative stables, the general Anglophone public and its satellite constituencies (or its client societies, if you prefer) would feel certain that Tagore's writings amounted to a corpus of academically irrelevant material assembled by a well-meaning amateur, which could be safely set aside by serious workers today. This imaginary sequence of responses would unfold at a great distance from the concerns of the Tagore-Einstein conversations and their contemporary methodological sequels. We obviously don't want to embark on such a self-defeating course of action. That is why we have not chosen, in this chapter, to try and push Tagore's critique of Chatterji per se to some sort of logical conclusion specifically in terms of the empirical material Tagore drew our attention to.

What are we doing instead, and how is it supposed to help?

Point Two A (v)

Our strategy is as follows. We take a thematic, methodological leaf out of Tagore's book and indicate some lines along which a fresh critique of today's successor framework (thematically identical to the target of his criticisms) can be mounted with respect to empirical phenomena that the scholars targeted by such a critique have no basis for brushing aside as belonging to irrelevant non-core domains. We do this without prejudice to the independent need for all linguists concerned with Indian languages to flag the particular phenomena Tagore was interested in. As pointed out in note 4, there is still empirical work to be done in order to address certain obviously core-linguistic issues that Tagore noticed before any professional linguist did. However, if we were to regard those particular grammatical insights as the only take-away from his writings about language, we would be missing the point.

Point Two B

Tagore's main point in his writings about language in general and his gentle critique of Chatterji in his 1938 book in particular is that the naive Galilean method – as Einstein embodied it in his work and in his conversations with Tagore – is in principle ill-equipped to cope with any linguistic expression (any word or combination of words) that does not convey its content through the standard, arbitrary, articulate or referential meaning-bearing machinery. In particular, the naive Galilean method, represented by its lexico-grammatical implementation in linguistics, stops in its tracks when it meets any *expression* whose functioning crucially relies on perceptual *impressions* feeding the listener's understanding of what has been said. The impression in question is by no means an obscure *je ne sais quoi*, as readers prejudiced by the fact that Tagore was primarily a poet may assume. In any given instance, the impression is either based on the sound-imagery of an onomatopoeic word, or on doubling that appears to mimic some repetitive or extended phenomenon, or on some other pertinent facts that can be particularized but that do not in grammatical terms add up to a fully specifiable outcome.

It was far from Tagore's intentions to deny or undermine the pertinence or value either of the lexico-grammatical machinery in linguistics or of a Galilean account of the physical universe (both of which are conceptualized in the same spirit of explicit formal articulation envisaged as capable of exhaustively accounting for all observables). On the contrary, he was a lifelong personal friend of Jagadis Chandra Bose, and an ardent champion of the scientific method who often emphasized that it was imperative for public figures like Gandhi to take science seriously. Tagore's point was that the *naive* Galilean method should not be accepted as a *comprehensive* approach either to the universe or to language, for *impressions* have to be allowed to count, and to count even in the final analysis, not merely at the level of a person's subjective preparation for a serious reckoning. Tagore argued that it was inappropriate to imagine that serious, ultimate answers would have to be impression-free in the interests of 'objectivity.'

Today, the debate between Tagore's side and Einstein's side over a viable version of the programme of Galilean inquiry takes a much sharper form. It is appropriate in our times to speak of a Galileo-Chomsky research programme in linguistics, which invokes the canons of science in general and physics in particular as a valid inspiration for lexico-grammatical inquiry. The question of the adequacy of this programme can be sharpened by taking a thematic leaf out of Tagore's book: in this chapter, we pose the matter in terms of the choice between a *sensitive* Galilean programme, informed by semiotics and thus capable of taking impressions on board by paradigmatic means, and the *naive* Galileo-Chomsky programme. The overwhelming majority of scholars (usually called the mainstream) have been using the latter as a first approximation. Its empirical inadequacy should have become obvious by now both to allies and to critics. Alas, it has not; we must therefore make a case for the sensitive Galilean programme in general – specifically, in the present context, we shall make a case for supplementing formal methods in linguistics, in particular, with ancillary, semiotic resources.

My teacher Ray C. Dougherty, in his class lectures at New York University in 1978 – the lectures manuscript version remained unpublished because the mainstream, whose editors and referees control the journals, has been systematically hostile to all such approaches – pioneered a semiotics-sensitive (but otherwise Galilean) approach to grammatical analysis. In the context of studying conditional structures, he considered descriptions coordinate-conjoining the *if* and *then* clauses and those subordinating the *if* clause to the *then* clause. The well-formed conditionals, Dougherty argued, were the ones for which a coordinating as well as a subordinating description was available. But he noted that, crucially, the two types of analysis were not amenable to exact comparison, and that therefore one would have to perform what he termed 'gestalt comparisons.'

Dougherty's proposal is akin to the semiotics-laden formal architectures for grammatical theory that two other mentors of mine, Ruwet[16] and Kelkar,[17] proposed independently of each other (and of Dougherty). Their frameworks – deploying semiotic resources to supplement the incomplete toolkit of naive-Galilean syntax and morphology – provide an effective response to one pervasive fact about the spread of grammatical features across languages. I am referring to the fact that syntactic constructions in diffuse languages are often inexactly aligned with morphological counterparts in compact languages. The phenomena of such inexact alignment across the syntax-morphology boundary were catalogued especially carefully by Tesnière.[18] Inexact alignments look intractable only in a system with no well-defined place for them. The point, surely, is to tweak the architecture to solve the problem. In this connection it becomes natural to extend to semiotics the methodological proposal made by Tirumalesh[19] for pragmatics. Tirumalesh proposed that pragmatic descriptions of phenomena that resist mechanical, core-grammatical treatment should use suggestive 'signposts' instead of explicit rules or other rigour-maximizing formal devices.

I shall now briefly survey three sets of linguistic facts for which the naive Galilean approach fails what linguists call the 'descriptive adequacy test.' Adding semiotics to our toolkit spectacularly advances our understanding of the phenomena in all three contexts.

Point Two B (i)

Formal linguists postulate 'missing elements' all over the place, systematically extending an insight that users of language have always taken for granted. We all agree that there is a missing 'you' in an imperative sentence like *Stop writing now*; this can be technically called 'deletion of a designated element' – only a second person subject gets deleted in an imperative, which means the listener can uniquely infer the content of the missing subject. Likewise, we all agree that in the sentence *Ali was hoping to stop smoking, and he has indeed managed to* there is a missing verb phrase *to stop smoking* right after *managed to*; formal syntacticians call this 'deletion [of the second occurrence of the verb phrase] under identity [with its first occurrence].' On the basis of such phenomena, syntacticians have set up a 'principle of recoverability of deletion,' which claims that every missing element can be recovered because what is deleted either belongs to a unique basket of designated elements or is identical to a guarantor expression that stays in place elsewhere in the sentence.

In recent work[20] I have shown that certain apparent counterexamples to this principle cannot be handled by rigorous formal means. I argue that semiotic slippage can explain how the relevant Bangla sentences nonetheless come out well-formed.

The issue concerns cases like *ekhane æmon kono ujjɔl alo nei ja dekhe tomar cokh dhãdhiye jabe* ['There is no such bright light here, seeing which your eyes will dazzle']. The principle says it should be possible to delete the relative pronoun *ja* ['which'] thanks to the guarantor *æmon kono ujjɔl alo* ['any such bright light'], but that the non-finite verb *dekhe* ['seeing'] has no right to go missing. However, one variant of the sentence appears to delete them both the relative pronoun and the verb: *tomar*[21] *cokh dhãdhiye jabe æmon kono ujjɔl alo ekhane nei*: lit. 'your eyes will dazzle, there is no such bright light here.' There is no known way for formal syntactic resources to address this apparent violation of the otherwise sacrosanct principle of recoverability of deletion.

I have argued in the syntax paper mentioned in note 20 that a semiotic slippage allows a shift from the relative to the complement clause construction, as in *ekhane æmon kono ujjɔl alo nei je tomar cokh dhãdhiye jabe*: 'there is no such bright light here that your eyes will dazzle.' On that assumption, that the troublesome variant arises through innocuous deletion of the complementizer particle *je* 'that,' which is a designated element. As I pointed out in my earlier paper on Tagore's contributions to syntax (see the Visvabharati Quarterly reference in note 4), Tagore made some useful remarks about affinities between relative clauses and the complementizer particle *je*. The current discussion thus directly reconnects with a cluster of problems that he flagged in 1938.

Point Two B (ii)

In another paper,[22] I take up another thread from Tagore's collectanea – the theme of word doubling, which linguists call reduplication. In that paper, too, I argue for a semiotics-dependent solution that remedies the inability of the standard formal machinery of grammatical theory to address an empirical problem in Bangla. The crux of the problem appears where the adverbial participle construction overlaps with infinitival self-embedding. The adverbial participle doubles the infinitival 'verb-***te***' form. You see these basic facts illustrated in the adverbial participle for which I use bold type in *raka* ***kãdte kãdte*** *cole gælo*: 'Raka went away ***crying***,' to be compared with the single word infinitival in *ami kɔkhono rakake* ***kãdte*** *dekhini*: 'I've never seen Raka ***cry***'].

In the light of these facts, consider a self-embedded infinitival construction like "raka dilipke bonnar kɔtha *bolte* **dite** caYni": 'Raka didn't want to **let** Dilip *talk* about the flood' (I am using bold type and italics in such examples in order to help you to match the right English gloss with the right infinitive in the Bangla original, and I employ double quotes for the original sentences and single quotes to mark the English gloss).

Adding a further twist, now consider a case where an infinitival embeds an identical infinitival, as in "raka dilipke bonnar kɔtha *bolte* **bolte** caYni": 'Raka didn't want to **tell** Dilip to *talk* about the flood.' Such a sentence, in writing, looks unparsable at first sight. When you intone it in a way that dramatizes the intended reading, a listener understands you but finds the sentence avoidably contorted, though grammatical. The empirical task is to answer the question 'what accounts for this impression? why do such sentences sound contorted?'

The answer to this question has something to do with the sharp contrast one observes between one-word infinitival verbs like *kãdte* 'to cry,' *bolte* 'to tell/talk,' and two-word verbs like *baron korte* 'lit. prohibition to do, = to prohibit.' It turns out that an infinitival verb that consists of two words can freely self-embed any infinitival, even if the embedder and the embedded are identical. Thus, one doesn't get a 'contorted sentence' impression at all when one hears or reads a sentence like "bibeker cɔmokprodo bɔhorer bokami dekheo dilip oke tobu Sɔb bæparei *baron korte* **baron korte** caYni": 'even in the face of the surprising scale of Bibek's silliness, Dilip still didn't feel like absolutely **forbidding** him to do any *forbidding*.' Here, crucially, the embedder verb **baron korte** and the identical embedded verb *baron korte* are both two-word sequences. Consequently, the **four**-word sequence of identical infinitives is sharply distinct from the **three**-word sequence we find in the corresponding adverbial participle construction, namely _*baron korte korte*_, as in "dilip kichu khete cailei tumi baron korcho; tumi oke Somane *baron korte korte* klanto hoYe jabe na to?": 'whenever Dilip feels like eating something, you are forbidding him; you won't get tired constantly *forbidding* him, will you?'

While a general reader may accept this material as evidence establishing the necessity of a resort to semiotics, many linguist readers will not. For their sake, I need to also state the adjectival half of the reasoning in order to secure the argument.

There is a similar phenomenon in the realm of Bangla adjectives. Nominal constructions in Bangla frequently mark plurality not by inflecting the noun, but by doubling the adjective, e.g. "Sɔccho Sɔccho camoc": 'lit. transparent transparent spoon, = transparent spoons.' Now, consider a self-embedded adjectival construction like "mukeS [kɔYekTa *Sɔccho* [**Sɔccho** plasTike moRa] camoc] kineche." Both in this Bangla original and in its English gloss 'Mukesh has bought [a couple of *transparent* spoons[23] [wrapped in **transparent** plastic]],' I am using square brackets to show how one adjectival construction is embedded within another; you also find the two kinds of typographic highlighting familiar from the verb cases. The upshot is that this adjectival case of identical self-embedding has a contorted sound to it, just as in the verbal case, and for the same reason.

The moment you add the intensifier word 'very' and turn the adjectival *word* into an adjectival *phrase*, the effect of contortion disappears, just as you would expect from what you learnt in the verbal case parallel to this one: "mukeS [kɔYekTa ottonto *Sɔccho* [ottonto **Sɔccho** plasTike moRa] camoc] kineche," 'Mukesh has bought [a couple of very *transparent* spoons [wrapped in very **transparent** plastic]].' Again, the way the language uses adjective doubling to affirm nominal plurality in these cases is to double the adjectival word proper, not to double the two-word adjectival expression as a whole: "ottonto Sɔccho Sɔccho camoc," 'very transparent spoons.' Thus, cases of adjectival construction self-embedding that involve two-word adjectival phrases do not come out contorted, as that particular kind of self-embedding cannot possibly be confused with the plurality marking device that doubles the word.

At this point, a defender of the naive Galilean approach is going to exclaim, rightly, that it is a matter not of semiotics but of grammar that adjective doubling marks plurality and that infinitive doubling flags adverbial participle status. Of course. To stop at this observation, however, is to fail to see where the semiotics kicks in. First of all, as Tagore notes throughout his linguistic writings, such doubling achieves its effects by appearing to mimic facts about the external world. An adverbial participle denotes a continuous action, and verb doubling counts as a gesture saying, look, the action is prolonged. A plural nominal expression using adjective doubling uses it to say, I'm not pointing at one thing, I'm hand-waving at a bunch of things. In no language does word doubling ever mark the singular rather than the plural. Second, we might add, it is a semiotic fact that no language uses a device of *phrase* doubling to mark participiality, plurality or any other grammatical status; what is doubled is always a word or a fraction of a word. It is the business of semiotics, not grammar, to delimit the word as a unit of linguistics – a point we cannot afford to labour when non-specialists are patiently waiting for the chapter to end.

Point Two B (iii)

As if it was not bad enough for the naive Galilean view of grammar that its wall has been breached at nouns, at verbs and at the fundamental conservation principle, we are about to find that even morphological agreement between inflected words is compelled to do business with semiotics, leaving no semiotics-free zone in any serious subdomain of grammatical theory.

I take it that even readers innocent of linguistics have heard of Saussurean arbitrariness – the idea that the same animal is known as a 'cat' in English and a 'beral' in Bangla for no good reason, as a matter of arbitrary convention. Let's call this level of arbitrariness 'proto-lexical' because it is a foundational fact about the lexicon of every language. We shall distinguish it from the additional 'deutero-lexical arbitrariness,' DLA for short, that we observe in languages like German. In DLA languages, grammatical genders are 'extra arbitrarily' associated with words for inanimate objects – German has *der Löffel* 'the spoon (masculine),' *die Gabel* 'the fork (feminine),' *das Messer* 'the knife (neuter).'

Now, what does it take for a language to have the DLA property? To put it differently, what features of its agreement system correlate with this characteristic? My stand is[24] that the following generalization holds in this domain: *An agreement language featuring gender exhibits deutero-lexical arbitrariness if and only if the language shows (a) gender agreement somewhere in the syntactic system and (b) agreement (not necessarily involving gender) between the noun and some other (usually a demonstrative) constituent of the nominal.* That some specialized writings designate agreement within the nominal construction as 'concord' rather than 'agreement' is a wrinkle we shall ignore.

If this generalization stands, we conclude that a language goes DLA wherever the grammatical architecture of the language sends gender into the agreement bloodstream instead of just leaving it on the skin, on the perceptible surface. This, again, is a feature of the grammatical system that grammar cannot, on its own, account for – just as grammar cannot delimit the word as a unit. It is the semiotics that is responsible for these deepest facts of architectural geometry shaping what is or is not possible in the grammatical architectures of particular languages.

Concluding methodological remarks

Right before I conclude, I need to address item (z) flagged under Point Two A (iii) above. I said there that I would need to convince you *that the widespread belief (which you quite probably share) that a methodological abyss separates historical linguistics from the natural sciences (an abyss that renders Tagore's critique of Chatterji obsolete because linguistics has long moved away from the spot he is targeting), is ill-founded.* It is now time to try.

I have argued at some length, elsewhere[1] – in an article that uses the term 'anarchic material' to label what in the present context can be described as material of Tagorean interest, and uses the term 'kernel' for the residue of a language that one obtains after one has subtracted all anarchic material from it,

> that the old logic of diachronic reconstruction has been driving the modern chariot of synchronic description as well. What linguists call "descriptive generalizations" – the basis of many truly admirable achievements of contemporary linguistics, but a basis that has long awaited serious clarification – mimic diachronic reconstructions of proto-forms. The anarchic lexical material methodologically excluded by descent-focused and contact-focused comparativists is excluded not just initially, but throughout. It remains outside the pale of systematic linguistic description even at the synchronic level. Though we pretend to describe complete languages, actually we have only been describing kernels – the generous kernels of contact studies rather than the austere ones of descent studies, to be sure, but still kernels. That we sometimes add a list of anarchic items as a postscript is a bit of a cruel joke.[2]

Readers inclined to contest my claim that no methodological abyss separates classical historical linguistics from the consequences of applying the naive Galilean method to the synchronic study of

1 My paper 'Situating 1786 in the History of Linguistics', presented at the Asiatic Society's seminar on '225 years of the German translation of Kālidāsa's drama *Abhijñānaśakuntala* and the science of comparative philology' on December 7, 2016; text accessed on January 22, 2018 at www.academia.edu/30274194/Situating_1786_in_the_history_of_linguistics

2 Ibid, 7.

languages are welcome to engage with the paper I just quoted from, and to refute its reasoning in detail, a controversy best pursued elsewhere.

Notes

1 Bertrand Russell, *The ABC of Relativity*, New York: New American Library, 1958 (4th printing 1962), pp. 124–5.

2 Rabindranath Tagore, *Naibedya* ['oblation'], Kolkata: Adi Brahmasamaj Press, 1901 [1308 Bengali Era, hereinafter 'B.E.'].

3 A serious unpacking of such thoughts, along with citations from the poems and from Debendranath, is provided in my article 'Fear and Beauty in Tagore's *Naibedya*', *Jadavpur Journal of Comparative Literature*, 2011, 48: 93–114.

4 I myself have done such mining work in two articles: 'The Rethinking of Language', *Visvabharati Quarterly*, 1985, 48(1–4): 138–50; Rephrasing the Question of Complex Predicates in Bangla: A Biaxial Approach, in Rajendra Singh and Shishir Bhattacharja (eds), *Annual Review of South Asian Languages and Linguistics 2012*, Berlin and New York: Mouton de Gruyter, 2012, pp. 3–44 (see especially section 4.1).

5 The passage that follows, in my rough and ready English rendering, is from section 13 (specifically, pp. 423–4) of that 1938 book as it appears in the standard edition of the collected works of Rabindranath [Rabindranath Tagore, Bāṅlābhāṣā-paricay, in his *Rabindra-Rachanabali*, vol. 26, Kolkata: Visva Bharati, 1949 (Paus 1355 B.E.), repr 1978 (Paus 1384 B.E.), pp. 367–456]. In this article I use standard Romanization and transcription systems; in the Bangla examples cited, T D R are retroflex, Y W are mid glides, S is palato-alveolar, c j are prepalatal affricates, and other symbols are as in the International Phonetic Alphabet. Tagore presents Sanskrit forms with erudite Bangla in mind, writing *vikaśita* 'blossomed', for instance, with a palatal *ś*; I have substituted the dental *s*, more usual in Sanskrit, to avoid stopping Sanskrit-proficient readers for irrelevant reasons.

6 The unfamiliar symbols here, like æ and ɔ, are drawn from the International Phonetic Alphabet.

7 It is standard practice to insert dots within a word, and within its gloss, in order to indicate boundaries delimiting what are called 'morphemes', i.e. significant parts of the word. To see how the notation works, consider the word *un.ity* and its paraphrase or gloss 'one.ness'. In this example, the special-purpose dot separates the element *un* 'one' from the affix *ity* 'ness'.

8 These approximate renderings within brackets are intended to assist readers who have no access to Bangla. They are not seriously equivalent to the expressive elements: as Tagore notes, ordinary meaningful words never achieve such equivalence.

9 Rabindranath Tagore, Śabdatattva [first published in 1909 as a collection of articles that had appeared in periodicals between 1885 and 1904], in his *Rabindra-Rachanabali*, vol. 12, Kolkata: Visva Bharati, 1949 [Paus 1355 B.E.], repr. 1978 [Paus 1384 B.E.], pp. 335–410.

10 This chapter of *Śabdatattva* first appeared in the journal *Sāhitya Pariṣat Patrikā* in 1900. Bibliographic information regarding individual chapters is provided on p. 630 of vol. 12 of his collected works cited earlier.

11 Suniti Kumar Chatterji, *The Origin and Development of the Bengali Language*, Kolkata: Calcutta University Press, 1926.

12 Available in English translation as: Rabindranath Tagore, *Letters from Java: Rabindranath Tagore's Tour of South-East Asia 1927*, ed. Supriya Roy, trans. Supriya Roy and Indiradevi Chaudhurani, Kolkata: Visva Bharati, 2010.

13 Available in English translation as: Suniti Kumar Chatterji, *Indonesia: Travels with Tagore*, trans. Rimli Sengupta, New Delhi: Indira Gandhi National Centre for the Arts, 2016.

14 The preface, bearing the title 'Bhūmikā' ('preface') and the subtitle 'Chātrapāṭhakder prati' ('to student readers'), appears in vol. 26 of the collected works (see note 5 for full reference), pp 369–71. We find the relevant sentence on p 371: 'This book *Viśvaparicay* which I wrote for you was also written along the same lines' – here Tagore uses the familiar form of 'you', *tomader*, the form normally used by a teacher speaking to students.

15 Rabindranath Tagore, Viśvaparicay, in his *Rabindra-Rachanabali*, vol. 25, Kolkata: Visva Bharati, 1949 [Paus 1355 B.E.], repr. 1978 [Paus 1384 B.E.], pp. 345–415. Available also in English translation: Rabindranath Tagore, *Our Universe*, trans. Indu Dutt, Mumbai: Jaico, 1999.

16 Nicolas Ruwet, *Syntax and Human Experience*, trans. and ed. John Goldsmith, Chicago: Chicago University Press, 1988.

17 Ashok Ramchandra Kelkar, *Language in Semiotic Perspective: The Architecture of a Marathi Sentence*, Pune: Shubhada Saraswat, 1997.

18 Lucien Tesnière, *Eléments de syntaxe structurale*, Paris: Klincksieck, 1959.

19 Kilingar Venkappa Tirumalesh, 'Why You Can't Repunch Bill: An Inquiry into the Pragmatics of *Re*-Words in English', *Journal of Pragmatics*, 1991, 16: 249–67.

20 Probal Dasgupta, 'Pre-Demonstrative Gaps in Bangla: Syntactic and Semiotic Recoverability', *Język. Komunikacja. Informacja*, 2016, 11: 195–212.

21 I need to warn non-technical readers that the initial 't' of this sentence is compelled to remain lower-case, in defiance of ordinary typography, on account of specialized notational conventions that reserve capital T, D, R for retroflex sounds. For the sake of aesthetic balance we use lower-case initials in all such Bangla examples.

22 Probal Dasgupta, 'Getting the Identical Infinitives Filter in Bangla Under Control', in Gautam Sengupta, Shruti Sircar, Madhavi Gayatri Raman, Rahul Balusu (eds), *Perspectives on the Architecture and Acquisition of Syntax: Essays in Honour of R. Amritavalli*, New York: Springer, 2018.

23 The bewildered non-speaker of Bangla needs to be told that an unmarked noun can count as plural without any grammatical device indicating its status; contextual cues are enough. What the adjective doubling in the earlier example does is to unambiguously specify a plural interpretation.

24 See my 2015 Suniti Kumar Chatterji Memorial Lecture delivered at the Indira Gandhi National Centre for the Arts on November 16, 2015 – text accessed on January 22, 2018 at www.academia.edu/18501933/Classify_or_Agree_The_View_from_a_Split_Classifier_Language. The details provided in the present article are emended findings, first made public at my Leiden University talk in 2016.

References

Chatterji, Suniti Kumar. 1940. *Indonesia: Travels with Tagore*, trans. Rimli Sengupta, New Delhi: Indira Gandhi National Centre for the Arts, 2016.

———. 1926. *The Origin and Development of the Bengali Language*, Kolkata: Calcutta University Press.

Dasgupta, Probal. 2018. Getting the Identical Infinitives Filter in Bangla under Control. In Gautam Sengupta et al. (eds), *Perspectives on the Architecture and Acquisition of Syntax: Essays in Honour of R. Amritavalli*, New York: Springer.

———. 2016. *Situating 1786 in the History of Linguistics*. Unpublished conference presentation. Text accessed on January 22, 2018 at www.academia.edu/30274194/Situating_1786_in_the_history_of_linguistics

———. 2016. Pre-Demonstrative Gaps in Bangla: Syntactic and Semiotic Recoverability. *Język. Komunikacja. Informacja*, 11.

———. 2015. *Classify or Agree: The View from a Split Classifier Language*. Unpublished conference presentation. Text accessed on January 22, 2018 at www.academia.edu/18501933/Classify_or_Agree_The_View_from_a_Split_Classifier_Language

———. 2012. *Rephrasing the Question of Complex Predicates in Bangla: A Biaxial Approach, Annual Review of South Asian Languages and Linguistics 2012*, Berlin and New York: Mouton de Gruyter.

———. 2011. *Fear and Beauty in Tagore's* Naibedya. *Jadavpur Journal of Comparative Literature*, 48.

———. 1985. The Rethinking of Language. *Visvabharati Quarterly*, 48.

Kelkar, Ashok Ramchandra. 1997. *Language in Semiotic Perspective: The Architecture of a Marathi Sentence*, Pune: Shubhada Saraswat.

Russell, Bertrand. 1958. *The ABC of Relativity*, New York: New American Library.

Ruwet, Nicolas. 1988. *Syntax and Human Experience*, Chicago: Chicago University Press.

Tagore, Rabindranath. 1938. *Bāṅlābhāṣā-paricay*. In *Rabindra-Rachanabali*, 26, Kolkata: Visvabharati, 1978.

———. 1937. *Viśvaparicay*. In *Rabindra-Rachanabali*, 25, Visvabharati, 1978; available in English as *Our Universe*, trans. Indu Dutt, Mumbai: Jaico, 1999.

———. 1935. *Letters from Java: Rabindranath Tagore's Tour of South-East Asia 1927*, trans. Supriya Roy and Indiradevi Chaudhurani, Kolkata: Visvabharati, 2010.

———. 1909. *Śabdatattva*. In *Rabindra-Rachanabali*, 12, Kolkata: Visvabharati, 1978.

———. 1901. *Naibedya*, Kolkata: Adi Brahmasamaj Press.

Tesnière, Lucien. 1959. *Eléments de syntaxe structurale*, Paris: Klincksieck.

Tirumalesh, Kilingar Venkappa. 1991. Why You Can't Repunch Bill: An Inquiry into the Pragmatics of Re-Words in English. *Journal of Pragmatics*, 16.

3

TAGORE'S TRUTH

Amiya Dev[*]

I shall approach the issue in hand, namely, the nature of reality apropos the Tagore-Einstein dialogue, as a student of literature. For, to me, Tagore is primarily a creative artist. His poems and songs, his plays, his musical and dance dramas, stories and novels, essays and discourses, autobiographies and travelogues, even his letters are where his truth abounds, more than in his 'philosophical' writings in English. I have been rereading some of those above these past four or five years; and what follows is my search for Tagore's truth in them. I beg my reader's indulgence for my inept English renderings – except in one case – of the Bengali originals I have quoted.

Let us begin with his novel *Gora*, a classic by now. Laid out in the second half of the nineteenth century, its 76+ chapters raise a number of issues, those of the Hindu versus the Brahmo, the 'superstitious' versus the 'enlightened,' the tolerant versus the intolerant, the British ruler versus the Indian ruled, compliance cum subservience versus non-compliance cum freedom of spirit, self-acclaimed and exclusive Hindutva versus all-embracing Indianness. It is Tagore's 'India' text par excellence. For Tagore, love of one's motherland is not biological, but a conscious act. We may like to read/hear his batch of songs called 'Svades' (Motherland) along with *Gora*, where too the relation with the motherland is non-compulsive, analogous to the realization of true 'motherhood' in the novel. But what happens if that love is manipulated as shown in *Ghare-Baire* (*The Home and the World*), his first political novel built around the Boycott Movement as an aftermath to the Anti-Bengal Partition upsurge. The slogan *Bande Mataram* becomes a travesty though not without causing Hindu-Muslim tensions. The truth that is made home is the need for a better selfhood (*atmasakti*, as developed in a series of essays under that title) to qualify for the love of motherland. Mere opposition to the foreign ruler is not enough. When we get to his second political novel, *Car Adhyay* (Four

Chapters), written in the background of *Svadesi* (nationalist terrorism), that truth takes on a new meaning: under no circumstances, no matter how politically opportune, must we sacrifice our inner humanity, our *svabhav* in other words. It is a tale of tragic waste in the name of patriotism.

Gora came out in 1910. But in 1910 also came out *Gitanjali* (An Offering of Songs),[1] followed four years later by *Gitimalya* (A Garland of Songs) and *Gitali* (A Gathering of Songs), the three forming a cluster. The truth in them is the truth of a 'you' beyond the poet's 'I,' and of their relationship, which is spiritual though devoid of any formality. Nothing could possibly be more intense. Tagore's fame, at least one-time fame as a mystic, is largely based on that. This truth is affirmed by his play *Raja* (*The King of the Dark Chamber*), which is also a 1910 title (though it actually came out in early January 1911) – any hint thereof of an *annus mirabilis*? Based on the Kuśa Jataka (where physical ugliness was the reason for King Kuśa hiding from his queen yet performing wondrous deeds for her) it has an interplay of the visible and the invisible, of form (*rup*) and the formless (*arup*), reportedly giving the philosopher Wittgenstein a new perception. Not far from this batch is *Dakghar* (*The Post Office*), the play whose truth too has two levels, of death and awakening (death to 'unbelievers' and awakening to 'believers'), supplemented by a call to the faraway symbolized, as it were, by the letter from the king (the title of André Gide's translation *Amal et la lettre du roi* highlights that). It made history by being staged in Polish in July 1942 by a Jewish orphanage in the Warsaw Ghetto under Janusz Korczak's guidance, perhaps to gather courage before transportation to Treblinka. It was also broadcast by the Paris radio on the eve of the Nazi occupation of Paris on 14 June 1940.

The Post Office is dated 1912, which also saw the appearance of the play *Acalayatan* (Immovable Precincts), dealing with a different truth, the ossification of an institution through dead habit and blind exclusiveness, which desperately needed to be overhauled to bring about inclusiveness and freedom. But the demolition is to be followed by new growth, although that too may be demolished if the new goes the way of the old (any hint of 'perpetual revolution' minus its political steam-rolling?). Two later plays, *Muktadhara* and *Raktakarabi* (*The Waterfall* and *Red Oleanders* in Tagore's own translation) also contain similar human truths. Telling the story of a dam that was built to cut off water to people downstream to cow down their spirit of independence, *Muktadhara* is acutely relevant to today's dam-led crisis causing large-scale displacement and deprivation. On the other hand,

Raktakarabi is about industrial labour's alienation and dehumanization. Not only has industry lured farmers away from fields and fitted them into its cobwebs, but has also turned them into numbers. This is ahead of Wordsworthian Romanticism where industry merely meant the rape of nature.

Going back to his first major play *Bisarjan* (*Sacrifice*), what do we get there? Religious idolatry in which the temple priest thrives, enraged by the liberal king's orders to stop live sacrifice to the goddess. To maintain his authority – not out of faith though – he demanded royal blood for sacrifice. He is jolted back to his senses when his foster son whom he had raised without knowing his *kshatriya* origins and whom he loved next to his life, offered his own blood to the idol. That sacrifice helps establish the human truth of fatherly emotions. If, along with this, we read Tagore's *Santiniketan* sermons mainly given to his pupils at the *Brahmacaryasram*,[2] we will probably see that his critique of idolatry has no hint of atheism or agnosticism. He does imbibe in them a sense of the divine. Here is an excerpt from the one titled 'Prem' (Love) given in the middle of December 1908:

> In the *Vedasuktas* it is said that death is His reflection, that deathlessness too is His reflection – He has turned them both into one. He in whom has all conflict been resolved is the ultimate truth. He is the purest light, He is the clearest darkness.
>
> If all the contraries of creation are not united in one truth, then that cannot be taken as the ultimate truth. If for what is left over from that, another truth has to be made up, then these two are to be treated as one another's opposites. Then God is to be taken for deathlessness and Satan for death.
>
> But we do not believe in any partner of the Brahman – we know that He is the truth itself, that all the conflicts of part-truths have been resolved in Him. We know that He is one; that all the part-truths' differences have been united in Him.
>
> But this is only a theory. That He is the truth is only an apprehension in knowledge – what is its relation to our hearts? Cannot this truth be relished?
>
> But how can we say that? As all truths have merged in Him, so have all *rasas* merged in Him. Hence the Upanishads have not only called Him the truth but also the essential relish – knowing Him as that fully laid-out *rasa* is perfect knowledge.
>
> So it transpires that He who is the ultimate truth is also the absolute *rasa*. That is, He is the essential love. Else nothing could have been resolved in Him; difference would have only

> stayed difference, conflict would only have been hurting and death would have only been snatching away. Everything uniting in Him is not a union per knowledge – there is an aspect of love in it – that is why everything has to be united, why separation and conflict can never turn into a true thing.

If we turn now to his novel *Jogajog* (Relations), we encounter another truth in a clash of values between old-time dignity and *nouveaux-riche* arrogance. Here marriage clashes with money revealing woman's vulnerability in pregnancy. Yet it is not a simple tale of maladjustment; there is a hint of frustrated relationship that gives the truth its complexity. Of course a need is felt for woman's self-determination, the clearer instance of which has been given in the story *Strir Patra* (Wife's Letter) with a wife walking out of her husband's life and her family in-law. In the same vein is *Payla Nambar* (House Number One) where the wife walks out on both husband and lover. His earlier stories, especially the ones written during his stay in riverine Eastern Bengal looking after the family estates, are a close portrayal of rural life in its multiplicity of experience, including marriage and dowry, property disputes, relationships and the joint family upheavals, with a special focus on women. The truth that transpires from them may simply be paraphrased as *samsara* or the life here and now (his use of the word '*samsarjatra*' or 'living the life here and now' in a way encrypts that). But there are also stories of individual destinies of various kinds, of love and hope of fulfilment, love and separation, of the tragedy of love's triangle as shown in the long story *Nashtanid* (Broken Nest). Speaking of love, his dance drama *Syama* (towards the close of his life) reconfirms the very tragedy embedded in love, the last stanza by the male lead being a prayer for *kshama* or mercy for not being able to pardon the female lead's cruel sacrifice, for his sake, of her young unhappy lover, too eager to be the scapegoat.

And of course, in search of his truth, I have been looking at Tagore's poems and songs, composed all his life holding a whole range of experience, from the least tremor in the psyche to the most wondrous perception of reality, doubts and questionings not overruled. His songs being a perfect union of word and melody have a special place in his works – they are worth relishing again and again, their truth often hitting us as an epiphany. It is almost like what he has said in a song beginning 'The night my doors were smashed in a storm': '*in the morning I looked and saw you standing in my roomful of emptiness, what wonder!*' I have also been looking at his self-portraits, done in his last decade – past *The Golden Book of Tagore* (1931), that memorable celebration of him on his seventieth birthday – and been trying to fathom their truth. What sadness! What a piercing self-search that denies any Olympian tranquillity!

2

Are these truths transitory and various not amounting to a unity? That I suppose is the moot question for me. Since I would not deduce from what Tagore told Einstein and fit it in with his creativity, my task is simple and may even be a little impressionistic. For to me Tagore is not the philosopher to question Einstein's science, but a creative genius of the highest order whose lifelong adventure with word, rhythm and melody and in late life, with line and colour may give him access to a *Weltanschauung*, but no system. Knowing his roots in the Upanishads, can we work our way back to them from his creative output, giving him the credentials, as it were, of a latter-day *Brahmavadin*? (To my mind, this would be less likely than tracing the *Commedia* back to Thomism.) An instance that comes quite close in this regard is the novel *Caturanga*'s (Quartet) Sachis whose quest for the absolute (we remember him rushing out into the dark once saying aloud 'You are mine, O You Infinite, You are my own!')[3] comes up against his pull to Damini. Branding her *Prakriti* (mindless nature) does not really help, for she is too full of flash (true indeed to her name, lightning) to be cold-stored as a symbol. What he exhorts earlier about the formless (*asim*) 'forever descending into form' thus inducing form (*sima*) to 'run after the formless,' is not a cold abstraction spelling *lila*, but touched by the passion of a seeker not unaware of form's snares. The famous cave episode is obviously more than a nymph's lure: such temptation is verily a two-way experience. Tagore's truth here has antecedents in his tradition.

To go ahead with my lay reading, let me take a cue from Goethe, who called his autobiography written in the prime of his life *Aus meinem Leben: Dichtung und Wahrheit*, thus combining poetry with truth.[4] Although Tagore's first autobiography, also written in the prime of his life, is simply called *Jibansmriti* (*My Reminiscences*), he too is concerned with poetry *and* truth. It begins with his discovery of rhyme in his childhood, the lilt, that is, of words. What follows is an account of his formative years, focusing on a keen sensibility and a corresponding inscape, yielding full-throated juvenilia, exercise upon exercise, till poetry answered his call. Indeed the answer came accompanied by his perceptions of the 'forest-like heart' and his release thereof. The latter takes the shape of an epiphany, the famous episode of sunrays flooding one morning from behind the neighbourhood trees, while he was standing on the balcony of the Sudder Street house in Central Kolkata where he was staying at the time with his elder brother. That this had the spirit of a truth is also evident from an essay called '*Manabsatya*' (Human Truth) appended to *Manusher Dharma* (Human Religion), the Bengali companion, three years later, to his Hibbert Lectures in 1930:

> The moment the sun appeared from behind the trees, the veil from over my mind was lifted. I felt I was seeing the truth in the naked eye. I was looking at man's inner self.

Another truth of such import is sounded in *My Reminiscences* while, speaking of the 'unfathomable darkness' pulling him during his first real bereavement at the age of 24, he also records 'a sudden breeze of joy blowing now and then' in his mind. This may take us to a prose poem written some years later:

> At dawn she left us.
> My mind began consoling me, 'All is illusion.'
> I said angrily, 'But look, there is her sewing kit on the table, flower pots on the terrace, her autographed fan on the bed – isn't all that real?'
> The mind said, 'Still consider – '
> I said, 'You keep quiet. Look at the book of stories, a hairpin stuck at a page in the middle, she hadn't finished reading it; if this too is illusory, then why had she to be more illusory?'
> The mind kept its silence. A friend came and said, 'What is good is true, it never vanishes; the whole world adorns it like a gem pendant from its necklace.'
> I said angrily, 'How do you know? Is the body not good? Where has that body gone?'
> As the little boy hits his mother out of anger so did I keep hitting at all around that sustained me; I kept muttering, 'This world is faithless.'
> Suddenly I was startled. Someone seemed to be saying, 'Ungrateful!'
> I looked out of the window and saw the crescent moon rising behind the fern, as if the one who had gone was playing hide and seek. From the star-strewn darkness came a scolding, 'Was my being willingly there so false, and this parting so terribly true?'
> ('*Kritaghna Sok*' [Ungrateful Mourning], *Lipika*, 1922)[5]

While *My Reminiscences* flash one set of truths, another set is flashed by his letters to a niece (1887–1895) from Eastern Bengal – letters cum diary – where he had his full encounter with nature including man in nature's proximity. These were a prelude to a good many of his poems, in colour rich and in craft innovative, but more, as said above, to a whole batch of short stories, of both lyrical and dramatic flavour. The letters, *Chinnapatra* or loose leaves, are often reflective and may have an indirect bearing on his dialogue with Einstein. Here is an excerpt:

> I have often thought that the profound joy we have from nature is because of our feeling of an enormous relationship with her – these ever-reviving green and sappy grass and creepers, trees and bushes, this flow of waters, this blowing of winds, this continuous revolving of the shadows, this cycle of seasons, this ever-moving stream of stars and planets filling up the boundless sky, the endless kingdom of life on earth – with all this is our blood-stream in the arteries linked – we are set to the same rhythm as the whole universe – where this rhythm is taking a pause, or where it is ringing out, there from inside our heart comes a consent – if all the particles and atoms of nature not our kin, not ever throbbing in our heart out of a sense of beauty and profound joy, then we would never have had such inner happiness to be in contact with the world outside us.
>
> (23 August 1895)

This is unity and man alone is capable of such a perception. 'The bird you have given song,' said Tagore in a poem in 1915 (*Balaka* 28), 'it sings that song,/it doesn't give more./Me you have given voice, I give more,/ I sing.' The same is said of wind and the full moon – they don't give more than they have been given. Man alone does that. The poem ends on:

> To others you only give,
> from me you take.
> I offer you love,
> smiling you step down from your throne
> to clasp it to your heart.
> You get back
> more than you gave.

God or no God to whom this is addressed, it is a paean to man. We may also recall that familiar song on Song, '*akas bhara*' (1924), where the key word is 'wonder', which man alone can have:

> Sky-full of sun and stars, a cosmos-full of life,
> in the midst of that I have found my place,
> *of wonder then is my song born.*
> Waves of endless time that rock the earth in flood and ebb
> have given a pull to my arteries' blood,
> *of wonder then is* my *song born.*
> I have stepped on grass on the way to woods,
> by flowers' fragrance has my heart wakened,
> gifts of joy are strewn around,
> *of wonder then is my song born.*
> I have laid my ears, opened my eyes, poured my heart into
> earth's bosom,
> among the known I have searched the unknown,
> *of wonder then is my song born.* (emphasis added)

This may take us back to an earlier song (1897) around 'wonder', though of a somewhat austere tone:

> In this universe, these greater skies, in eternal time
> I alone, man, go about in wonder, go about in wonder.
> You lord of this universe, in an unending mystery plunged
> keep to silence, alone, in your own abode of glory.
> In this boundless space and time, among these countless worlds
> of light,
> your gaze is upon me – I keep gazing at you.
> All commotion suspended, all creation in peace –
> One you are, in you I am, alone, without fear.

This I suppose is the closest he could come to Einstein's universe. Mystery is highlighted, an approach held unthinkable, one may even imagine a little of the fear that Pascal once spoke of ('Le silence éternel de ces espaces infinis m'effraie.' – *Pensées*, I. 91). Yet, to whatever extent, the subjectivity is there, and that makes the difference. Now, can we not give the Tagore-Einstein dialogue the benefit of the doubt and say that Tagore and Einstein had their different orientation to truth, which they could not compromise? It may indeed be idle to tilt the balance either this way or that way. Not that Tagore was totally unaware of the impenetrable. In *Manusher Dharma*, he did say in Chapter 2:

> There is no point in calling the impersonal Universal Being that goes entirely beyond the human being, desirable or anything for that matter. He is bereft of any distinction between good and bad, beauty and ugliness. As regards our relation to Him we cannot speak of vice or virtue. . . . He *is* – that is all we can say of Him. We have heard that it is possible to immerse in that Non-Particular by giving up all aspects of human consciousness. There cannot be any arguing that. Within the parameter of the human mind, how can we ascertain if one has gone beyond one's being along with the mind itself? . . . There could be a mind somewhere the world perceived by whom is beyond our mathematical computation, which does not belong to the *akasa* (spaces) we call ours. But how can we call the world whose secrets man is deciphering by his mental faculties, superhuman?

But is Einstein's impersonal universe 'superhuman' in this sense? I cannot claim any confidence here and leave it to scientists. Let them tell us if Einstein's 'Realism' leaves no room at all for human perception, even Pascal's kind of fear. All I can do at this point is quote a little further from Tagore:

> This world I know by my own perception. The agent that has this knowledge is my *soul*; he also knows himself. This self-revealing *soul* is not alone. My *soul*, your *soul*, his *soul*, many such *souls* are there. The *soul* in which they attain truth is called *oversoul* in our scriptures. This *oversoul* is *human-oversoul*. He is *sadā janānāṁ hṛdaye sanniviṣṭaḥ*, He is ever present in all human hearts.

And in his preface to *Manusher Dharma* Tagore calls him

> Universal and Everlasting Man. It is by his pull that universality appears in man's thought, feeling and action. Great men easily realize him in all mankind and easily sacrifice their life for love of him. It is through a perception of him that man crosses his life-limits and arrives at the doors of Universal Man.

Perhaps we have a key here to Tagore's emphasis on human consciousness in understanding reality. Of course, it is not an individual man's consciousness, but consciousness at the level of Universal Man. He simply cannot think of a universe outside his Universal Man's view. We can recall a good many songs he composed at various times of his life to illustrate this, obviously irrespective of his dialogue with Einstein. In one such song he uses the image of a shepherd grazing scores of suns and stars in the ever-wide fields of heaven, with no greater diligence than tending grasses and weeds and bushes on earth.

3

But to take these as Tagore's final words on reality would be denying him his subsequent creativity. I am not saying that he relinquishes the human consciousness; that would be perjury after that poem 'Ami' (I) in *Syamali*, his book of poems dated 1936:

> It is from my consciousness' colour that the emerald grew green,
> the ruby red.
> I looked up at the sky,
> light flashed
> from east to west.
> Looking at the rose I said, beautiful,
> it became beautiful.

Yet the most astute philosopher would think twice before confirming this as Tagore's 'truth' and on that count, celebrating his abiding calm. Tagore's long day's night was too anguished to allow him that calm. Goethe is said to have ended his life with a wish for more light – we remember Tagore quoting this in one of his *Chinnapatra* letters dated 22 September 1894 and saying that under similar circumstances he would have asked for 'more light and more space,' the immediate motivation being his love of the Bengal landscape. But what was his last wish? We are told that towards the very end, from his semiconscious state, he wanted to hear his own words from *Gitanjali's* Song No. 4: '*bipade more raksha karo/e nahe mor prarthana,/bipade ami na jena kari bhay*' (Save me from peril/is not my prayer,/let me be not afraid of peril). Had he made his peace with reality? Yes, but with much disillusion about its ways. Creation, he said, was full of wiles, with snares of false beliefs laid for great hearts. It is a tortuous path often landing one in ridicule. Truth he finds in his own light, and if he can bear the guiles to the end he wins the lasting right to peace. In a preceding poem, dictated one late night three months before his death, Tagore demystified truth. It is hard, he said, but once borne it can be cherished. It does not delude. The last three lines of that poem, stark to their marrow, deny any calm whatsoever:

> Up to death is life a meditation of suffering,
> to win the terrible value of truth,
> to pay up all debts in dying.

And a month ago in his last birthday speech (*Sabhyatar Samkat* or *Crisis in Civilization*) he had said that it would be a sin to lose faith in man. Do we not also remember that he had a song added when its Bengali original was read out beginning '*oi mahamanab ase*' (There the saviour comes)?

I suppose the difference between a philosopher and a creative artist is marked when we recognize that a little over a month before that 'I' poem in celebration of human consciousness, Tagore had painted that striking self-portrait where his fire-coloured face streams down with two penetrating but somewhat sad eyes, face flanked by two darkish spaces – a haunting image (see below).

In no ostensible way can we relate the painting to the poem. And within months of the poem he wrote out that unusual piece of narration called *Se* (He: 1937) where the logic of the subject is taken back to story-telling itself. The archetypal story begins in 'Once upon a time there was a/an . . .' – why not just '*se*'? If the first person 'I' am telling the story, and the second person 'you' are listening, then who can it be better about than the third person '*se*'? And as such '*se*' does not have to have a name or time-place co-ordinates. This indeterminacy gives it a futurist character where the narratee and narrator are as crucial as the narrated. No less perhaps are his illustrations to *Se* – typical Tagore – some surely vie with those accompanying his nonsense rhymes in *Khapchada* (Disorderly).

Now, this is another area that defies a straight answer to Tagore's truth – I mean his paintings. Have we yet come to terms with them? Are they, as some say, expressions of his subconscious? Or are they, truly, expressions of that divine thirst for rhythm, which has been his, only in a new dress? The sheer bulk of them in such a short time is stupendous – and it was not that he had given a holiday to word and melody, and every other preoccupation, above all Santiniketan for which he had to go 'begging.' Indeed Tagore *is* a phenomenon.

Tagore once said that one of his watermarks had been inconsistency. What he probably meant was not sticking to one truth or one

২৪ এপ্রিল ১৯৩৬-এ অঙ্কিত কবির আত্মপ্রতিকৃতি

Figure 3.1 Self-portrait of Tagore, dated 24 April 1936

Source: Courtesy of Rabindra Bhavana, Santiniketan, West Bengal

set of truths in his creations of art but coming up with a whole variety of truths. Whether or not they add up, our immediate concern, with the Tagore-Einstein dialogue in view, is the means by which these truths are mined. Tagore's world is a world of sense experience, not of abstract speculation. His grains of sand revealing an eternity are not an eternity walking its way back to a grain of sand. In other words, his world cannot be *objective* of its own without even the veiled presence of his *subjectivity*. And is not this the way of art in general, as opposed perhaps to that of science? Einstein's violin or S. N. Bose's *israj*, for that matter, could have been intriguing, if its need had not been felt. Yet Einstein's $E = mc^2$ or S. N. Bose's 'statistics' was in no way jeopardized by the violin or the *israj*. (Let us not also overlook the Tagore-Einstein dialogue on music.) If they had not been overly preoccupied with themselves in 1930, then Tagore's belief in human truth might have made sense to Einstein and Einstein's belief in non-human truth to Tagore. For he did express in one of his last poems, in tune with the two I have spoken of, a sense of bafflement at the Being:

> The first day's sun
> asked
> at Being's new appearance –
> who are you,
> no answer came.
> Years went by,
> the day's last sun
> put the last question by western seas
> in silent evening –
> who are you,
> had no answer.

Today Tagore is an open book to us, its last pages as valid as the first. We can taste it anywhere at any moment and have a touch of the truth therein. By truth I do not mean the absolute but the perception that an instance of beauty is lined with. Obviously there are hundreds of such instances. Do the accompanying perceptions cohere? In other words, do we arrive at *one* truth from Tagore? If that involves exclusiveness, putting one set aside to advance another, we may demur. Surely, Tagore's truth is not a singular one in the sense of Gandhi's truth, which is a value, a moral drive involving his life and action. Yet, surely again, Tagore's truth has a singularity in its human perspective. His last days' bafflement is as humanly felt as his early or middle days' wonder and inwardness. A creative artist cannot but be humanly stirred – as much in his song offerings as in his doubts and questionings. The purely non-human may eventually be the scientist's domain. The objective world he analyses is independent of him; it exists in spite of him. As such he may not have any lack of subjectivity that may even influence his method – apropos serendipity we are at times told of a random choice. But by no means can he make a dent on the objectivity of his specimens. The apple falls.

Notes

* When I had presented (or perhaps should say performed, for a live presentation is indeed a kind of performance) a version of this chapter at the three-day conference at the Indian Institute of Advanced Study, Shimla in 2012 on 'The Nature of Reality: The Perennial Question', I had it dedicated to my late friend Jaidev who had been my immediate association to Shimla. He had taught English at the Himachal Pradesh University, the Indian Institute's neighbour on the Summer Hill. But before long he had realized that teaching English in India would not be a contradiction in terms if he could relate to the life around and to the word produced on the Indian soil. His conscientiousness had eventually brought him to Mahasweta Devi's works. He had even made it once to her tribal festival (*Sabar Mela*) in Purulia. At his death 19 years ago Mahasweta Devi had dedicated two of her stories to him (see Pankaj K. Singh (ed), *The Politics of Literary Theory and Representation*, in memory of Jaidev, Manohar, New Delhi, 2003). I would like to keep to my dedication.

1 Lest there be confusion to someone not familiar with Tagore's Bengali oeuvre, this is not his English *Gitanjali*: *Song Offerings* that fetched him the Nobel Prize and gave him international fame without which there might not have been an occasion for his dialogue with Einstein on the nature of reality. Of course, the Bengali *Gitanjali* contributed a bulk of his translations forming the *Song Offerings*, 53 out of 103.

2 These 152 sermons cum meditations cum prayers are stretched between December 1908 and December 1914. See *Santiniketan* in *Rabindra-Racanabali* (Collected Works of Rabindranath Tagore), 150 Year Commemoration Edition, Vol. 16 (Kolkata: Pascimbanga Bangla Akademi, 2016), pp. 172–173.

3 This and the following two quotes are in Asok Mitra's translation (see *Chatunranga*, New Delhi: Sahitya Akademi, 1963, pp. 88, 87).

4 If Plato had indeed denied truth to poetry, then Aristotle surely had reinstated it.

5 There is another poem in the same batch of the same import, called '*Pratham Sok*' (First Mourning).

4

SEEING THINGS

Tagore's sense of the real

Supriya Chaudhuri

Reality is the most contested term in human discourse, and it is not surprising that the poet and the physicist could not agree on its nature. Instead of attempting to resolve their differences, therefore, I will begin by opening up a distance from science and philosophy, and look, not so much at the nature of reality, as at the literary techniques of 'realism.' Fortunately, I have the poet to keep me company, and I will start with a close reading of Rabindranath Tagore's essay 'Bastab' [The Real], first published in the journal *Sabuj Patra* in 1914 and later reprinted as the opening chapter of his book *Sahityer Pathe* in 1936.[1] Tagore proposes here a critical examination of literary realism, seeing 'the real' as an ideological construction, not as a given of human perception. I will argue that in his fiction, no less than his poetry, he offers a critique of the simple identification of 'the real' with the things of the world. My reading will draw examples from Tagore's prose writings, rather than his poetry, attempting to draw out a debate on 'realism,' rather than a metaphysical questioning of 'the nature of reality.' I will suggest, however, that the disagreement between Einstein and Tagore in their conversations of July 1930 turns on the impossibility of reconciling the physical with the metaphysical.

Seeing things

In *A Vision of the Last Judgment* (1810), William Blake famously declared:

> What it will be questioned When the sun rises do you not see a round disk of fire somewhat like a Guinea O no no I see an Innumerable company of the Heavenly host crying Holy Holy Holy is the Lord God Almighty I question not my Corporeal or Vegetative Eye any more than I would question a Window concerning a Sight I look thro it & not with it.[2]

The apparently demented confidence of the poet as visionary challenges orthodox notions of perception and representation, but there can be no doubt that Blake was very much aware of the claim he was making. There is no comparable statement in Tagore, but he had, I would suggest, an equally shrewd awareness of the impossibility of seeing 'the thing itself' – the sun, in this case – unmediated by the monetary calculation that views it as 'a round object somewhat like a guinea,' or the madness that views it as an innumerable company of the heavenly host. In his later years he found himself increasingly answerable to critics who complained about the lack of realism in his work, and in his 1914 essay, 'The Real,' he offered a set of extended meditations, not so much on the real, as on the way the real is understood. In many ways this is a puzzling essay, addressing the question of the real, as of realism, tangentially rather than directly, and it has not to the best of my knowledge been translated, though several other essays in *Sahityer Pathe*, including 'Sahitya' ['Literature'], 'Kabir Kaiphiyat' ['The Poet's Defence'] 'Adhunik Kabya' ['Modern Poetry'] and 'Sahityatattva' ['The Philosophy of Literature'] were admirably translated by Swapan Chakravorty about ten years ago, enabling wider access to a debate about realism in Bengali literature that acquired urgency in the second decade of the twentieth century and was continued, in the context of the aesthetics of modernism, in Bengali literary periodicals of the 1920s and 1930s.[3]

The essay on 'The Real' is to some extent an answer to the attack launched by Bipinchandra Pal in the journal *Bangadarshan* in 1912,[4] complaining about the lack of materialism (*bastutantra*) and likeness to reality (*bastab*) in Rabindranath's works: a clearly political diatribe that Rabindranath confronts by repeatedly deferring its object of enquiry – the real – to questions of belief, of ideology, and of the nature of literary representation. As commentators have noted, the debate, especially as taken forward by other critics like the Marxist social historian Binoy Ghosh, comes to focus on issues of social class and elitist bias, the idea of history and the idea of the nation. In his chapter on 'Nation and Imagination' in *Provincializing Europe*, Dipesh Chakrabarty argues that 'realism and historicism came together in this debate as the twin pillars supporting the ideas of democracy and historical materialism.'[5] Rabindranath's adversaries felt that the ancestral wealth and protected upbringing of the Tagore family had cushioned him against the harshest of everyday realities. Rabindranath asserted, by contrast, that he had depicted the life of Bengal's forgotten villages in his short stories more faithfully and with closer knowledge than any of his middle-class critics.[6]

Prose and poetry

Nevertheless, the issue of how reality was to be represented was an unavoidable one. Chakrabarty sees the question of realism as in effect creating a fissure between Tagore's poetry (*kabya*) and his prose (*gadya*), though this 'division of labour,' as he calls it, is much less absolute than he suggests. In fact Chakrabarty's complex, historically elaborated argument requires a kind of prose-poetry binary that buys into some of the key positions in the debate between Tagore and some of his younger critics in the 1920s. Yet, as he himself notes, these positions were always-already undermined by the shifting categories (and registers) of the poetic and the prosaic, both in Tagore's own work and in the poetry of his successors. Lyric universalism is never absent from Tagore's 'seeing' of the world, both in the letters of *Chhinnapatrabali* [Scattered Leaves] and in the prose fiction of *Galpa-guchchha* [A Bunch of Stories], the two works most directly influenced by his contact with Bengal village life (apart from the essays he wrote for his efforts at rural reconstruction around Shantiniketan), while the *remaking* of the poetic register is central to the anti-Tagorean modernist aesthetics of the 1920s. The terms 'poetry' and 'prose' function as place-holders for a range of affective and social concerns, but not in a single and unchanging way. Chakrabarty's argument aligns the 'real' on the side of what he calls 'historicism' (that is, the representation of temporally located and verifiable material settings), and with the politics of democracy and nation-building, while the poetic 'imagination' appears an ahistorical interloper, constantly threatening to dissolve the painful realities of poverty, injustice and time-bound suffering. Yet even in the highly deliberate and resonant applications of the terms 'poetry,' 'prose' and 'reality' that Tagore and others make in the course of this debate, it is always obvious that the terms are interchangeable. As Chakrabarty himself notes, Tagore lays claim to a truth beyond the external 'veil of reality,' while the poet Sukanta Bhattacharya's radical 'dismissal' of poetry against the background of the 1943 famine acquires potency precisely because it uses poetry to make its point:

> Poetry, I grant you leave today.
> In hunger's kingdom the world is suffused with prose,
> The full moon is like a scorched *chapati*.[7]

As Tagore sees it, both here and in the more famous essay on 'Modern Poetry' originally published in the journal *Parichay* in 1932, perceptions of the real in literature have almost everything to do with the ideology of the viewer. He cites, ironically, 'a critic's' reaction to his novel *Gora* (1910):

> A critic says that the only one of my works to contain some elements of realism is my novel *Gora*. What that work contains is least apparent to its author. People say that there is a good analysis of traditional Hinduism in it. From this I deduce that this must be the element of realism.[8]

But, as Tagore noted repeatedly, *Gora* is not a realist novel in the conventional sense: in fact it is a novel that rubs allegory, the instrument of poetic making, against the grain of history and the technical apparatus of realist representation as used in the high bourgeois novel. Gora is as much a symbolic figure as he is a realistic character, and the novel is like a fable of nativity, ironically setting the details of Gora's birth against his self-identification with the Hindu nation, thereby dismantling the latter. The point about the perspective of the viewer that Tagore makes in his comments here is further extended in the essay on 'Modern Poetry,' where he ends by describing both excessive respect for the world-as-object and unbidden hostility to it as *sentimentalism* (he uses the English word). In a sense, Tagore is going back to Friedrich Schiller's famous categorization in his essay 'On Naïve and Sentimental Poetry' (1795–1796), of modern, that is post-classical, writing, as sentimental.[9] Yet, of all the moderns, Tagore is surely the poet whom Schiller would have described – had he lived a century after to read and understand his work – as naïve, in the same sense as that by which Goethe to him was an example of naïve genius. For the modernist, by contrast, as Robert Musil noted in his diary entries for the composition of *Young Törless*, 'the reality one is describing is always only a pretext.'[10]

'What is the Real?'

Let us attempt to tease out some of the strands of this complex literary argument. In the first place we have the term *bastab*, which, as Tagore demonstrates in his rhetorical play on words in this essay, is derived from the word *bastu*, which may be translated as 'thing,' or perhaps as 'matter.' Then there is the term *bastabata*, which might mean 'reality,' or possibly 'realism,' though a stricter term for the latter would be *bastabbad*. Discussing *bastab*, or the real, as 'matter,' Rabindranath indicates the space of disagreement about what might constitute both the matter of poetry and the matter of the world. He says unambiguously that the matter of poetry is *rasa*, a central term in Sanskrit aesthetics variously translatable as 'savour,' 'sap' or 'juice':

> The point now is, what matter do we seek in literature? Experts tell us that the matter we seek is *rasa*.[11]

Who, he asks, can be the judge of such matter but the *rasik*? And who is to say who that *rasik* is? In fact, quite apart from Rabindranath's ironic puzzlement as to where the *rasik* might be found, the term *rasik* itself does not easily yield to translation. The *rasik* is the 'man of taste,' a person gifted with an aesthetic sensibility, or perhaps the Lord himself, alone capable of savouring the taste of the world, in so far as *rasa* means 'savour' or 'sap.' The question of literature (or of aesthetics) is thus thrown back upon a judgement of taste, and this Kantian dilemma is not one that we can easily escape. It is a problem that occupies Tagore both in this essay and the one that follows it in *Sahityer Pathe*, 'The Poet's Defence' ('Kabir Kaiphiyat'), since as he says in another essay, 'Soundaryabodh' ['The Sense of Beauty'], the first of the extension lectures he delivered at the National Council of Education, Bengal, in 1907, literature must ceaselessly express the dictum of the *Taittiriya Upanishad*: *raso vai sah*, 'He himself is *rasa*.'[12] For the poet as for the ordinary human subject, this *rasa* is certainly real: as the matter of emotion, or of feeling, it is the stuff of daily experience, and as legitimate an object of artistic representation as any other object in the world. Yet there can be no agreement as to whether *rasa* is represented or expressed, or whether it is the means of expression itself.

What do objects hold?

However, as Tagore grants, *rasa* resides in a receptacle, that is, it must be conveyed through a representation of objects or actions in the world, phenomena that may appear to us to have a more solid existence. It is in the choice of these objects or actions that he locates an ideological disagreement between the Romantic poetry he read in his youth, the Victorian 'realism' that surrounded him in his maturity, and the modernist dissidence by which he was assailed in his later years. The nature that he found in the poetry of Wordsworth and Shelley, or indeed the erotically charged landscape, as spiritual as it was physical, of the Vaishnava poets of Bengal whom he imitated in his first poetic productions, was for Tagore as much a body of real phenomena as the powerful banality of everyday life represented in the novels of Flaubert or George Eliot. And although the simple naming of material objects might be the source of a 'reality effect,' as Roland Barthes described it in his famous essay of that name, discussing a short story by Flaubert,[13] the techniques of realism are as grounded in worldly objects being suffused with feeling, as on such objects being present in and for themselves. The notation of the external features of a setting is never more than synecdoche, a fragment or shadow of full representation: the work that description does in the realist novel is dependent on what objects mean.

We may recall Erich Auerbach's seminal analysis, in his great book *Mimesis: The Representation of Reality in Western Literature*, of a scene from the first part of Flaubert's *Madame Bovary*:

> But it was above all at mealtimes that she could bear it no longer, in that little room on the ground floor, with the smoking stove, the creaking door, the oozing walls, the damp floor-tiles; all the bitterness of life seemed to be served to her on her plate, and, with the steam from the boiled beef, there rose from the depths of her soul other exhalations as it were of disgust.[14]

Auerbach asserts that though the paragraph presents a picture, that picture 'is not presented in and for itself; it is subordinated to the dominant subject, Emma's despair.' The reader sees the scene through Emma: 'directly, he sees only Emma's inner state; he sees what goes on at the meal indirectly, from within her state, in the light of her perception . . . the gloomy room, the commonplace food, the lack of a tablecloth, the hopelessness of it all.'[15] So although the passage is almost evenly balanced between material and immaterial referents (the boiled beef and the bitterness of life), the banality of the former is directly produced by the sensibility that experiences the latter. For this reason, Flaubert can slip effortlessly from the materialism of object description to the 'uncanny' (*unheimliche*) realm of metaphor: 'Her life was as cold as an attic facing north, and ennui, the silent spider, was weaving its web in the darkness, in every corner of her heart.'[16] Such description lends itself naturally to phenomenological analysis – as in Georges Poulet's brilliant reflections on time in Flaubert – because it is so clearly *intended*, as consciousness is intended, inclined or directed towards the represented object.[17] Auerbach sees this as an effect of style, the 'systematic and objective seriousness from which things themselves speak and, according to their value, classify themselves before the reader as tragic or comic, or in most cases unobtrusively as both,' and he defines 'the serious treatment of everyday reality' as the foundation of modern realism and of the democratizing impact of the novel.[18]

Realism and modernism

Tagore was capable of such description, as evidenced in his early short stories and in his later fiction, as well as in the letters of *Chhinnapatrabali*. Before moving on to a discussion of representative passages, however, we might note the counterbalancing, in these essays, of the Romantic themes of nature, beauty and art – illustrated, in another essay, 'Tathya o satya' ['Fact and Truth'], from John Keats's 'Ode on a Grecian Urn'[19] – against high bourgeois realism on the one hand, and the modernist focus on urban decay and squalor on the other. For there is a distinction between the apparent substantiality of realist object description and the disintegrating gaze of modernism. This latter too, as Rabindranath notes in his essay on 'Modern Poetry,' proclaims itself as a kind of realism, distinct only in its preference for a certain class of objects as against others. It evokes – as, say, its characteristic *rasa*, or feeling – not the 'excessive respect for the world-as-object,' a kind of *sentimentalism* by which Tagore characterizes the bourgeois aesthetic of object description, but an 'unbidden hostility' towards, or a repudiation of, the world and its objects, which he calls by the same name – that is, *sentimentalism*.[20] If the mid-Victorian aesthetic had deferred to reality, the present time attempts, in Tagore's opinion, to demean it. Earlier in this essay, Tagore has described 'pure modernity,' as a dispassionate absorption, free of personal attachment' – a clear, calm, impersonal and undeluded 'seeing,' which he contrasts with the corrosive distaste expressed through modernist aesthetics. This latter form of realist anti-realism is illustrated by the celebrated passage on the frog and Apollo, and involves an extended reference to T. S. Eliot's *Preludes*, cited partly in English and partly in Rabindranath's own Bengali translation. The poem's affective fusion with its urban setting, and its masterly image of an 'infinitely gentle/Infinitely suffering thing' produce their own shock of linguistic assimilation as the poem draws to its translated close:

His soul stretched tight across the skies
That fade behind a city block
Or trampled by insistent feet
At four and five and six o'clock;
And short square fingers stuffing pipes,
And evening newspapers, and eyes
Assured of certain certainties,
The conscience of a blackened street
Impatient to asssume the world.

I am moved by fancies that are curled
Around these images, and cling:
The notion of some infinitely gentle,
Infinitely suffering thing.

Wipe your hand across your mouth, and laugh;
The worlds revolve like ancient women
Gathering fuel in vacant lots.[21]

Yet despite Tagore's explicit rejection in this essay of both the ethic of despair and the aesthetics of object description that he attributed to modernist poetry, there were unarticulated tensions in his own relation to modernism. Dipesh Chakrabarty offers a close reading of his great poem, 'Banshi' ['Flute'], to reveal a poet divided between the impulse to outdo the modernists at their own trade, and the necessity of penetrating through reality's veil to the heart of the impoverished clerk who has abandoned his betrothed in her village by the Dhaleshvari river, 'dressed in a Dhaka sari, vermilion in her hair.'[22] And it might be noted too that while in the earlier essay on 'The Real,' Tagore finds himself under ideological attack because his writing is too Westernized and therefore, according to purists, not rooted sufficiently in real – or perhaps traditional – Indian life, in the later essay on 'Modern Poetry' he locates both Victorian realism and modernist anti-realism in the West. In another essay, 'The Philosophy of Literature,' Rabindranath asserts that the human race has set itself the task of 'mitigating the exclusive dominance of the clutter of material things. It is clearing a space for its feelings.'[23] This distinction between the nominal thing and the feeling that gives it life and presence is a recurrent theme in Tagore's writing, just as he contrasts the joy (*ananda*) of artistic creation with the utility of the created object's material function.

Techniques of representation

But in addition to the matter of poetry (*rasa*) and its vehicles (objects and events), for both of which the term *bastu* could be employed, there is also the technique of description, and indeed realist object-description, as all analysts of nineteenth century fiction are aware, depends almost entirely on the rhetorical figure of synecdoche, whereby the 'insistent particulars' of everyday reality do duty for the entire phenomenal world they represent. Realism is as much a mode of description as it is an ideology of being-in-the-world, and there is a close connection between that ideology and the realist faith in particulars that allows parts to substitute for the whole. In classical bourgeois realism, there is a kind of confidence in the solidity and self-presence of the object itself (illustrated in Claude Levi-Strauss's brilliant close analysis of Francois Clouet's portrait of Elizabeth of Austria, reproduced in *The Savage Mind*),[24] which only comes under attack when photographic illusionism and modernist dissolutionism take over, in the early twentieth century, the space of representation. Tagore is well aware of this transition, and in his late fiction he shows himself as adept at realist object description as he is distrustful of its ideological investment in the object world. In a late novel, *Jogajog* [Relationships], he uses the techniques of realist object-description to devastating effect in building up a sense of the crude materialism of the Ghoshal household. Yet he must have been preparing for the radical break with naturalism manifested in the paintings he produced in the last quarter of his life.

In effect, Tagore can be viewed, throughout his long writing life, as engaging with the problem of how the real might be seen, and how it may be represented. How does the poet see, and more importantly, what does he see? At one level, the answer he gives is that suggested by Shelley's account of the poet in *Prometheus Unbound*:

> He will watch from dawn to gloom
> The lake-reflected sun illume
> The yellow bees in the ivy bloom
> Nor heed nor see, what things they be
> For from these create he can
> Forms more real than living man
> Nurslings of immortality.[25]

The poet must be attentive to reality, spend time and labour observing the *minutiae* of the created universe, yet he must also, intensely, ignore it. Both the vivid particulars of the yellow bees and their immortal forms have presence and reality for the Romantic poet, and Tagore often describes himself as a Romantic. At the same time, as he admits in poems like 'Aikatan' ('How Little I Know of This Great World') or 'Ora Kaj Kore' ('They Work'),[26] there is a kind of duty to the human world of work and its phenomenal embodiment that is captured in the ethic of realism, which, because it focuses on being in the world, must be impelled by an ideology of care. This concern is represented not just in the poetry but also in the realist and semi-realist fiction of his middle and later period. There is a kind of cautious experimentalism in the short stories and in a novella like *Nashtanir*, where feelings, in a sense, become objects. Even without an exhaustively itemized object world, we feel ourselves to be in the presence of an authentically detailed experiential universe, where metonymy constantly slips into metaphor:

> In this way, Charu dug a tunnel under the entire structure of her domestic tasks and duties, and in that unlighted silent darkness she built a temple of secret grief, adorned with the garlands of her tears. Neither her husband nor anyone else in the world had any claim there. That place was as secret as it was deep, as it was beloved. At its entrance she would abandon all the disguises of her household and enter in her unadorned true form, and when she left it, she would put on the mask again and present herself in the theatre of the world's work, laughter and conversation.[27]

The filter of consciousness

There is a similar passage in Tagore's much later novel *Jogajog* ('Relationships'), though there, and in his later fiction generally, Tagore shows himself willing to engage much more implacably and unrelentingly with the weight and material substance of the world. In *Jogajog*, as I have suggested in another paper on furniture,[28] Tagore takes to its logical culmination a tension, a kind of fault-line, between the necessity of grounding representation in objects, and a profound distrust of the object world, that develops throughout nineteenth-century Bengali fiction. As a novel, *Jogajog* employs to the fullest extent the technique of realist object description that Tagore had learnt from the high bourgeois novel and which is largely encoded through description of furniture and domestic interiors, 'the gaze of obedient objects' about which Baudrillard writes with such contempt.[29] Furniture carries an enormous weight of representation: the description, metonymically precise, is also metaphorically overdetermined, so that the account of Kumudini's bedroom is conveyed through a process of itemization, or of detailed display. But the novel's descriptions are permeated too by a sense of profound aversion from those objects, a rejection of the phenomenal adequacy of that world, which mirrors Kumudini's own revulsion from the commodity culture embraced by her husband Madhusudan Ghoshal. Instead, Tagore substitutes the reality of music and meditation, the reality of memory and longing, as a form of affective substitution for a reality grounded in things (or such things) alone. For his heroine Kumudini also sees things, while things see her, as does the silent Subha in an eponymous short story, sitting face to face, at noon, with the intimate immensity of nature:[30]

> From all this, Kumudini came away to Calcutta. It was like a great ocean, yet where could she find even a drop of water to quench her thirst? At home in the country, even the sky and air had a familiar appearance. The horizon of the village was bounded here by dense forest, there by sandbanks, by the river's course, the spire of the temple, the empty sprawling fields, the copses of wild tamarisks, the tow-path. With different strokes in diverse colours, these had encircled the sky in such a way as to make it a special sky, Kumudini's very own sky. The light of the sun, too, was a special light. The lake, the fields of crops, the cane-forests, the red-brown sails of the fishing-boats, the shiny leaves of young bamboo-stems in the bamboo-groves, the smooth dark green of the jackfruit-trees, the pale yellow sand of the river's farther bank – the sunlight had mingled with all of these to acquire a familiar hue. In Calcutta, that sky and that light she had known so long, shattered into fragments by the hard unyielding lines of unfamiliar roofs and walls, stared at her harshly like a strange man.[31]

The poet and the scientist

Finally, and rather belatedly, I would like to take these reflections on the real and its representations back to the metaphysical concern that is, despite my own neglect of it, central to our collective deliberations. When the poet met the scientist, on 14 and 19 July 1930, they disagreed not so much on ways of seeing things as on the things themselves (though Tagore did use the analogy of the table that is seen and the paper that is written on, and Einstein spoke of the red flower on the table).[32] In the first day's conversation, where the disagreement was more strongly in evidence, Einstein offered a straightforward opposition:

> There are two different conceptions about the nature of the universe: (1) The world as a unity dependent on humanity. (2) The world as a reality independent of the human factor.[33]

That Tagore chose the first option had to do with his understanding of the terms *atman* and *brahman*, which have no precise equivalents in Western metaphysics, and are only poorly rendered by terms such as 'self,' 'consciousness,' 'divinity' or 'universal mind.' There is no necessary separation between *atman* and *brahman*: one finds realization in the other, and it is for this reason that Tagore says: 'The entire universe is linked up with us in a similar manner, it is a human universe.'[34] At a later point in the discussion we find him attempting to convey this notion by using the terms 'universal mind' and 'super-personal man':

Einstein: The problem begins whether Truth is independent of our consciousness.

Tagore: What we call truth lies in the rational harmony between the subjective and objective aspects of reality, both of which belong to the super-personal man.

Einstein: Even in our everyday life, we feel compelled to ascribe a reality independent of man to the objects we use. We do this to connect the experiences of our senses in a reasonable way. For instance, if nobody is in this house, yet that table remains where it is.

Tagore: Yes, it remains outside the individual mind but not the universal mind. The table which I perceive is perceptible by the same kind of consciousness which I possess.

The crucial point here as I understand it is contained in the last statement: 'the table which I perceive is perceptible by the same consciousness which I possess.' That is, in order for there to be a table *qua* table, the consciousness that perceives it as such must not only be shared by other human individuals, but it must exist *as consciousness*. For Tagore that consciousness, which perceives the table, is not restricted to the individual human but exists in what he calls the universal mind, somewhat like the Platonic idea of a table. So Tagore's insistence on the apparently subjective nature of reality, 'the world as a unity dependent on humanity,' is not based on a relativistic view of individual perception, a perception that differs from agent to agent, but upon the involvement of consciousness in the constitution of things as things. Einstein's choice of the second option ('the world as a reality independent of the human factor') is based on a narrower understanding of the *human* (in Western metaphysics, strictly differentiated from the *divine*); he insists that things exist in themselves whether or not there is a (human) consciousness to perceive them. Beauty, he concedes, may lie in the eye of the (human) beholder, but truth – the truth of things (and this is presumably both their physical *being* and the physical *laws* by which they are governed) – exists independently of human observation. For this argument regarding the physical reality of the universe, which Einstein provocatively calls his 'religion,' the question of whether things exist in the mind of God is irrelevant. Somewhat irreverently, we might recall at this point the celebrated exchange of limericks between Ronald Arbuthnot Knox – himself a Catholic priest – and an anonymous humourist:

There once was a man who said: 'God
Must think it exceedingly odd
If he finds that this tree
Continues to be
When there's no one about in the Quad.' *(Knox)*

Dear Sir, Your astonishment's odd:
I am always about in the Quad
And that's why the tree
Will continue to be,
Since observed by Yours faithfully, God. *(Anon.)*[35]

It is worth re-emphasizing, then, that the disagreement between Einstein and Tagore inheres not only in a difference of philosophical ground, but in a different understanding of the term 'human,' which for Einstein is a restricted category of individuals, while for Tagore it is a category that 'can only be realized by completely merging the individual in its infinity.'[36] No doubt it is fascinating to speculate that Tagore might have understood even then that the physical facts of object-existence, so far as we can describe them, depend on consciousness, and that this realization, by no means an original one at the time or even earlier, has been borne out by quantum mechanics. But in fact the dispute between the poet and the scientist is not on whether things exist in themselves, but on whether the physical 'facts' concerning them would remain facts independent of the human consciousness. Tagore's conviction that consciousness (for him as much a *universal* as the laws of matter) is involved in the very constitution of a physical truth makes him appear to be a spokesperson for the relativist and subjective view of reality. Not only does he recognize this problem, but towards the end of the first conversation, he states that there is an 'eternal conflict' between the universal human mind and '*the same mind confined in the individual*' (my italics).[37] That is, for him the question of relativism arises in the difference between universal and individual consciousness, a difference that needs to be overcome, given that the truth of things is ultimately a truth perceived by consciousness. He offers the following analogy:

> There is the reality of paper, infinitely different from the reality of literature. For the kind of mind possessed by the moth, which eats that paper, literature is absolutely non-existent, yet for Man's mind literature has a greater value of truth than the paper itself.[38]

The point is not that the paper does not exist as a material object or that it does not exist independent of moth or human. The point is that it exists in a certain way for the moth that consumes it as an edible object, and in a different way for the human mind that views it as a writing surface upon which signs are inscribed. The 'reality of literature' is non-existent for the moth but important for human beings. Although Tagore follows this up by saying that a truth that cannot be in some way grasped by the human mind remains as 'nothing,' it does not appear that he wishes to deny a mind to the moth, or reality to the paper independent of what is written on it. Einstein's exclamation 'Then I am more religious than you are!' suggests that he is worried by this interference (if we may call it that) from the human, since for him things and the truth of things exist according to physical laws that also govern human existence, and are independent of human consciousness of them. In effect what Einstein is rejecting is a form of Berkeleyean idealism, but it is very doubtful whether Tagore was a Berkeleyean idealist.

If the first day's discussion focused on reality – perceived or unperceived – the conversation on the second day begins with the relation between the individual and the general, freedom and law. Again, poet and scientist disagree not so much on the phenomena themselves, but on how they can be understood. For Einstein, causality is an absolute principle of the physical world, however disorderly the appearance of 'minute elements.' By contrast, Tagore chooses to postulate a metaphysical principle of harmony that controls the 'rebellious, dynamic' impulses of individual elements.[39] Given the distance between these two ways of viewing the world – physically or metaphysically – the exchange appears to have reached a kind of impasse. Tagore introduces a musical analogy, which Einstein seizes upon with relief, expressing polite curiosity about the difference between the prescriptive and regulated structures of Western classical music, and the free improvisations permitted to the Indian musician within the controlling melodic structure. Tagore makes several unavailing efforts to bring the discussion back to individual freedom and universal law, but it is quite clear that Einstein has realized the impossibility of explaining the laws of physics (as he understands them) in metaphysical terms. In consequence, they move on to a courteous exchange on musical practices east and west, and differences in taste and judgement. By the close, when Einstein offers the example of the red flower on the table that appears differently to different viewers, the discussion is no longer about the nature of reality *per se*, but about the nature of perception or of taste. Given that the flower is on the very table about which the two had disagreed in their earlier conversation, it is important to understand that Einstein has not actually yielded to a kind of subjective idealism here. As with the table, so with the flower – it would be on the table whether there is anyone to see it or not. Nevertheless, Einstein is prepared to concede that it 'may not be the same to you and me.'[40] By saying this he is accepting part of the argument from consciousness, but only that part that relates to the differences between individual human perceptions. The experience of viewing an object may vary from person to person.

Einstein is a physicist, not (or not necessarily) a physicalist, as philosophy uses the latter term. He does not, and indeed does not need to, take the further philosophical steps against physicalism famously argued for by Frank Jackson in his thought-experiment concerning the scientist Mary who had never seen the colour red, though she knew all there was to know about it.[41] According to Jackson, if Mary were to be let out of the black-and-white room in which she had spent her entire life collecting information in black-and-white, and if she were to be shown the red flower on the table, she would suddenly experience something new, the experience of colour *qualia*. Her experience might be quite different from those of other people who are accustomed to seeing red, and it might also be different from Tagore's own, given that he was probably colour-blind.[42] But even this philosophical thought-experiment does not turn on whether or not the flower (Jackson's example is that of a red tomato) exists when it is not seen by Mary, or on whether it does or does not possess the properties that lead most human eyes (not Tagore's, though) to view it as red. In fact as a scientist Mary knows these properties, though she has never seen a red flower. This notion of knowledge is also a subject of philosophical dispute, since, as Jackson argues, to experience the properties of things, their *qualia*, may also be regarded as a form of knowledge.[43] But Einstein is not offering a theory of knowledge here. Secure in his conviction that the flower and its physical properties exist independently of the presence of viewers, Einstein is willing to accept that the poet and the scientist view it differently. If pressed, he might even have suggested that consciousness has a physical basis. However, the conversation does not require him to go so far. On the other hand, Tagore, who has earlier stated unequivocally that consciousness is required for the act of constituting an object as object – therefore, the table as a table, the red flower as a red flower – is by this time more concerned to establish the relation between individual perception and a Kantian 'universal standard' which unceasingly attempts to reconcile differences. At this point, therefore – with respect to judgement or taste – Einstein appears to be relativist and particularist, while Tagore clings to the idea of a universal.

These shifts in position are deceptive. On issues of hard science – physics and cosmology above all – Einstein would not admit of differences linked to the human capacity for perception, and Tagore, however loyal to the notion of a universal principle, took the human as his point of departure. What was crucial, for him, was the *activity* of consciousness, its pursuit of abstraction in the midst of the vivid particulars of experience. For this reason, we cannot reconcile the scientist and the poet, since the first is concerned with knowing things, and the second with *how things are known*. Tagore's long endeavour as an artist and poet was to determine *how* things are seen, not *whether* they are there to be seen. Both scientist and poet would have agreed that there is a universe, and we are in it.

Notes

1 See Rabindranath Tagore, 'Bastab' ['The Real'], in *Rabindra Rachanabali* [Collected Works: hereafter *RR*], vol. 23, Kolkata: Visva Bharati, 1958, pp. 361–8.

2 William Blake, *The Complete Poetry and Prose of William Blake*, ed. David V. Erdman, Berkeley: University of California Press, 2008, pp. 565–6.

3 See Rabindranath Tagore, *RR* vol. 23, pp. 368–434, and the translations in Rabindranath Tagore, *Selected Writings on Literature and Language*, ed. Sukanta Chaudhuri, Sisir Kumar Das and Sankha Ghosh, New Delhi: Oxford University Press, 2001, pp. 258–309.

4 Bipinchandra Pal, 'Charitra-chitra: Rabindranath', *Bangadarshan*, Chaitra 1318 (March–April 1912), pp. 681–93. For the context of this attack see Prasanta Kumar Pal, *Rabijibani* [The Life of Rabindranath Tagore] vol. 6, Kolkata: Ananda Publishers, 1993, p. 301.

5 Dipesh Chakrabarty, *Provincializing Europe: Postcolonial Thought and Historical Difference*, Princeton: Princeton University Press, 2000, p. 157.

6 Rabindranath Tagore, 'Sahityabichar' (1940/41), in *Sahityer Svarup* [The True Form of Literature]: in *Rabindra Rachanabali* [Collected Works] vol. 27, Kolkata: Visva Bharati, 1956, p. 275.

7 Sukanta Bhattacharya, 'He Mahajiban' ['O great life'], lines 6–8, in *Chharpatra* [Letter of Leave], Kolkata: Sarasvat Library, 1955, p. 74. My translation.
8 *RR*, vol. 23, p. 364.
9 See Friedrich Schiller, 'On Naïve and Sentimental Poetry' (1795–96), trans. Julius A. Elias, in H. B. Nisbet (ed), *German Aesthetic and Literary Criticism: Winckelmann, Lessing, Hamann, Herder, Schiller, Goethe*, Cambridge: Cambridge University Press, 1985, pp. 180–232, esp. 190–5.
10 See J. P. Stern, '"Reality" in Early Twentieth-century German Literature', *Royal Institute of Philosophy Lecture Series*, 1983, 16 (September): 41–57, 42, for the reference to Robert Musil's diary entries in 1913.
11 'Bastab' ['The Real'] in RR vol 23, p. 362.
12 *Taittiriya Upanishad*, 2.7.2. See Rabindranath Tagore, 'Soundaryabodh', the first of the extension lectures he delivered at the National Council of Education, 13 Magh 1313 (January 27, 1907), published in *Sahitya*, 1907, and in *Rabindra Rachanabali* [Collected Works], vol. 8, Kolkata: Visva Bharati, 1941, pp. 355–72. Translated as 'The Sense of Beauty' by Swapan Chakravorty in Tagore, *Selected Writings on Literature and Language*, pp. 164–78, esp. p. 178.
13 Roland Barthes, 'The Reality Effect', in *The Rustle of Language*, trans. Richard Howard, Oxford: Basil Blackwell, 1986, pp. 141–8.
14 Cited from translation in Erich Auerbach, *Mimesis: The Representation of Reality in Western Literature*, trans. Willard R. Trask, Princeton: Princeton University Press, 1973, p. 483. See Gustave Flaubert, *Madame Bovary*, trans. Eleanor Marx Aveling and Paul de Man, ed. Margaret Cohen, London: W. W. Norton and Co., 2005, I.IX, p. 56, for a slightly different translation.
15 Auerbach, *Mimesis*, pp. 483–4.
16 Flaubert, *Madame Bovary* I.VII, p. 38.
17 Georges Poulet, 'Flaubert', in *Studies in Human Time*, trans. Elliott Coleman, Baltimore: Johns Hopkins University Press, 1956, pp. 248–61.
18 Auerbach, *Mimesis*, p. 491.
19 'Tathya o satya', *RR*, vol. 23, pp. 382–92.
20 See 'Modern Poetry', in *RR* vol. 23, pp. 420–34, translated by Swapan Chakravorty in Tagore, *Selected Writings on Literature and Language*, pp. 280–92, esp. p. 292.
21 Ibid., pp. 286–7.
22 See 'Banshi' (1932), from *Punascha*, *RR* vol. 16, pp. 84–7; translated as 'Flute-Music' by Sunetra Gupta in Rabindranath Tagore, *Selected Poems*, ed. Sukanta Chaudhuri, New Delhi: Oxford University Press, 2004, pp. 279–82. For Chakrabarty's discussion, see *Provincializing Europe*, pp. 164–7.
23 See 'Sahityatattva' in *RR* vol. 23, p. 437, as translated by Swapan Chakravorty in Tagore, *Selected Writings on Literature and Language*, pp. 293–309, esp. p. 296.
24 Claude Lévi-Strauss, *The Savage Mind*, [trans. Into English], London: Weidenfeld and Nicolson, 1973, illustration facing p. 148; discussion pp. 23–7 in 'The Science of the Concrete'.
25 Percy Bysshe Shelley, *Selected Poetry and Prose,* New York: Random House, 1951, p. 76.

26 For translations, see Tagore, *Selected Poems*, ed. Sukanta Chaudhuri, pp. 362–5, 368–70.
27 *Nashtanir*, chapter 15: in Rabindranath Tagore, *Rabindra Rachanabali*, vol. 22, Kolkata: Visva Bharati, 1957, p. 254. My translation.
28 See Supriya Chaudhuri, 'Phantasmagorias of the Interior: Furniture, Modernity and Early Bengali Fiction', *Journal of Victorian Culture* (Taylor and Francis) 2010, 15(2): 173–93.
29 Jean Baudrillard, 'Consumer Society', in Mark Poster (ed), *Jean Baudrillard: Selected Writings*, Stanford: Stanford University Press, 1988, p. 32.
30 See Rabindranath Tagore, 'Subha' (1893), in Sukanta Chaudhuri (ed), *Selected Short Stories*, New Delhi: Oxford University Press, 2000, pp. 105–6.
31 Cited from Rabindranath Tagore, *Jogajog/Relationships*, trans. Supriya Chaudhuri, New Delhi: Oxford University Press, 2006, pp. 52–3.
32 See 'Einstein and Tagore', in Sisir Kumar Das (ed), *The English Writings of Rabindranath Tagore*, vol. 3: *A Miscellany*, New Delhi: Sahitya Akademi, 1996, pp. 911–16.
33 *English Writings*, vol. 3, p. 911.
34 Ibid.
35 Cited from *The Oxford Dictionary of Quotations*, 2nd ed., London: Oxford University Press, 1972, pp. 305:10 (Knox) and 5:27 (Anon.).
36 Tagore, *English Writings*, vol. 3, p. 912.
37 Ibid., p. 913.
38 Ibid.
39 Ibid., pp. 913–14.
40 Ibid., p. 916.
41 See Frank Jackson, 'Epiphenomenal Qualia', *Philosophical Quarterly*, 1982, 32: 127–36, and Ibid. 'What Mary Didn't Know', *Journal of Philosophy*, 1986, 83: 291–5.
42 For this proposition, see Ketaki Kushari Dyson, *Ranger Rabindranath: Rabindranather sahitye o chitrakalay ranger byabahar*, Kolkata: Ananda Publishers, 1997, developing an idea first suggested by Stella Kramrisch, 'Form Elements in the Visual Work of Rabindranath Tagore', *Lalit Kala Contemporary*, 1962, 11 (December): 38.
43 See Jackson, 'Epiphenomenal Qualia', 128–30, 'What Mary Didn't Know', 291–4.

References

Auerbach, Erich. 1973. *Mimesis: The Representation of Reality in Western Literature*, trans. Willard R. Trask, Princeton: Princeton University Press.

Barthes, Roland. 1986. The Reality Effect. In *The Rustle of Language*, trans. Richard Howard, Oxford: Basil Blackwell.

Baudrillard, Jean. 1988. *Selected Writings*, ed. Mark Poster, Stanford: Stanford University Press.

Bhattacharya, Sukanta. 1955. *Chharpatra* [Letter of Leave], Kolkata: Sarasvat Library.

Blake, William Blake. 2008. *The Complete Poetry and Prose of William Blake*, ed. David V. Erdman, Berkeley: University of California Press.

Chakrabarty, Dipesh. 2000. *Provincializing Europe: Postcolonial Thought and Historical Difference*, Princeton: Princeton University Press.

Chaudhuri, Supriya. 2010. Phantasmagorias of the Interior: Furniture, Modernity and Early Bengali Fiction. *Journal of Victorian Culture*, 15(2): 173–93.

Flaubert, Gustave. 2005. *Madame Bovary*, trans. Eleanor Marx Aveling and Paul de Man, ed. Margaret Cohen, London: W. W. Norton and Co.

Jackson, Frank. 1986. What Mary Didn't Know. *Journal of Philosophy*, 83: 291–5.

———. 1982. Epiphenomenal Qualia. *Philosophical Quarterly*, 32: 127–36.

Kramrisch, Stella. 1962. Form Elements in the Visual Work of Rabindranath Tagore. *Lalit Kala Contemporary*, 11 (December).

Kushari Dyson, Ketaki. 1997. *Ranger Rabindranath: Rabindranather Sahitye o Chitrakalay Ranger Byabahar*, Kolkata: Ananda Publishers.

Lévi-Strauss, Claude. 1973. *The Savage Mind*, trans. into English, London: Weidenfeld and Nicolson.

Nisbet, H. B. (Ed.) 1985. *German Aesthetic and Literary Criticism: Winckelmann, Lessing, Hamann, Herder, Schiller, Goethe*. Cambridge: Cambridge University Press.

Oxford Dictionary of Quotations, The. 1972. 2nd ed, London: Oxford University Press.

Pal, Bipinchandra. 1912. Charitra-chitra: Rabindranath. *Bangadarshan*, Chaitra 1318 (March–April 1912), 681–93.

Pal, Prasanta Kumar. 1982–2003. *Rabijibani* [The Life of Rabindranath Tagore], vols. 9, Kolkata: Ananda Publishers.

Poulet, Georges. 1956. *Studies in Human Time*, trans. Elliott Coleman, Baltimore: Johns Hopkins University Press.

Shelley, Percy Bysshe. 1951. *Selected Poetry and Prose*, New York: Random House.

Stern, J. P. 1983. "Reality" in Early Twentieth-century German Literature. *Royal Institute of Philosophy Lecture Series*, 16 (September): 41–57.

Tagore, Rabindranath. 2006. *Jogajog/Relationships*, trans. Supriya Chaudhuri, New Delhi: Oxford University Press.

———. 2004. *Selected Poems* [in translation], ed. Sukanta Chaudhuri, New Delhi: Oxford University Press.

———. 2001. *Selected Writings on Literature and Language*, ed. Sukanta Chaudhuri, Sisir Kumar Das, and Sankha Ghosh, New Delhi: Oxford University Press.

———. 2000. *Selected Short Stories*, ed. Sukanta Chaudhuri, New Delhi: Oxford University Press.

———. 1939–2017. *Rabindra Rachanabali* [Complete Works], 33+2 vols, Kolkata: Visva Bharati.

———. 1994–2007. *The English Writings of Rabindranath Tagore*, vols. 4, ed. Sisir Kumar Das, New Delhi: Sahitya Akademi.

5

SCIENCE AND THE WORLD OF PERSONALITY

Shefali Moitra

In his conversation with Einstein, Tagore asserted, 'When our universe is in harmony with Man, the eternal, we know it as truth, we feel it as beauty.' Einstein's response was, 'This is a purely human conception of the universe.' Tagore replied, 'There can be no other conception. This world is a human world . . .' Tagore goes on to say, 'Science is concerned with that which is not confined to individuals; it is the impersonal human world of truths.' To which Einstein's reaction was, 'Truth, then, or Beauty, is not independent of man?' Tagore said, 'No.' Einstein said, 'I agree with regard to this conception of Beauty, but not with regard to Truth.' The present discussion will focus on these few excerpts from the July 1930 Einstein-Tagore dialogue.[1]

The conversation between Einstein and Tagore is commonly believed to be a nonstarter. The presumption being, they were talking to each other from two incommensurable paradigms – Tagore was speaking from the position of Upanishadic idealism and Einstein from the position of scientific realism. Thus, they were travelling on parallel paths. Further, it is presumed that since their areas of interest did not match, they were totally ignorant of each other's field of engagement – Tagore had a nodding acquaintance with science and Einstein had little acquaintance with the world of creative literature. We forget that Tagore had written a book on science[2] and Einstein wrote a poem titled, 'On Spinoza's Ethics.' Perhaps the general fate of all persons of genius is identical. They are all destined to be loners, they are not understood, and their views are seldom received in the proper perspective. What then was the outcome of their exchange? Did Tagore and Einstein remain worlds apart? The traditionally held interpretation is that they did uphold contrasting positions about the universe. I have my reservations on this point; there seems to be a considerable overlap between the two. In the present chapter I would like to show that Einstein's conceptual scheme could not have been worlds apart from that of Tagore – even though they were different. Given the opportunity of further conversation, many of their differences could have been ironed out and their compatibilities could have been brought to light.

From his childhood days Tagore was avidly interested in acquainting himself with recent developments in science. He was especially interested in discoveries in life science and physics. The interest was shared by other members of the Tagore family too. The poet's father, Debendranath took special care in disseminating contemporary scientific information to his children. Through an extensive exposure to the sciences Tagore could constantly steer clear of all forms of dogma, including religious dogmas. His deep understanding of physics, chemistry and life science is reflected in the analogies and metaphors he uses in his creative writings. Parimal Goswamy rightly points out that, though Tagore was not a scientist, he had the attitude of a scientist.[3] The presence of intellection can be found in all his works. His writings are very clear and well argued; there is no element of fuzziness or vagueness in his articulation. I agree with Goswamy, knowledge of science has facilitated the growth and formulation of Tagore's spiritual ideas. Goswamy goes to the extent of asserting, had Tagore not been exposed to the discipline of science, we would not have received him in his full form.[4]

Even though Tagore was a philosopher and not a scientist, questions relating to the nature of facts, the physical world, Truth, Reality and their relation to man always occupied an important place in his scheme of thought. This is evident from his *Sadhana* lectures delivered in 1913 to an audience in America.[5] The ideas contained therein were formulated much earlier through the discourses he gave to students in his school at Santiniketan. During Tagore's second visit to America he gave a series of lectures between September 1916 and January 1917. These were later published in a book form titled *Personality*.[6] In his second series of lectures the focus remained the same as that of the previous series. He revisits the topic of man's relation to the physical world and the spiritual world. This time he copiously cites the *Isha Upanishat* to further corroborate his ideas. The poet introduces his book *Sadhana* by saying, 'the subject-matter of the papers published in this book has not been philosophically treated. . . .'[7] Finally the same topic receives a more structured shape in his Hibbert Lectures delivered at Oxford and published with the title *The Religion of Man*.[8] The present conversation with Einstein took place in 1930, soon after Tagore delivered these lectures. We receive a mature account of Tagore's philosophy of man and his place in the universe in these lectures. These thoughts were also reflected in the poet's conversation with Einstein.

In his 'My Memories of Einstein' Tagore writes, 'My mind was full of the matter contained in my Hibbert Lectures which I had just delivered in Oxford.'[9] Tagore's sustained interest in the topics discussed with Einstein lead us to conclude that the conversation did not proceed on the basis of a series of impromptu responses to questions hitherto totally unattended by the poet. The poet himself was grappling with these issues for decades.

Tagore had categorically stated, he was not a philosopher, he was a poet. Tagore was both a poet and a philosopher. He was not an academic philosopher with a readymade meta-level theory for every first-level discipline. He had a conceptual scheme of his very own, which provided the standpoint from which he was conversing with Einstein. His conceptual scheme grew out of his first-hand experience. In his Preface to *Sadhana* he introduces his lectures as, 'my own special testimony, which must have its value because of its individuality.'[10] Being a creative individual his 'special testimony' is also unique. His thoughts cannot be neatly fitted into existing categories. Familiar terminology has been used with unfamiliar connotations. Tagore is a system builder. The system is a unified whole that cannot be factored into parts and studied in fragments. The key concepts used in his conceptual scheme are: fact, Reality, Truth, man, Man, freedom, reason, imagination, and personality. None of these concepts are standalone concepts for him, they are interconnected. A discussion of one concept necessitates a discussion of the entire gamut of concepts. Doubtless to say, giving an account of his position is a daunting task.

When Tagore, in his conversations with Einstein, says, 'This world is a human world . . .' he is indicating a special relation between 'this world' and 'humans'; similarly when he says, 'the universe is in harmony with Man, the eternal,' we need to dwell on the meaning he attributes to 'Man' (Man with a capital M), and to 'the eternal.' How does Tagore integrate the notions of 'man the mortal' with 'Man the eternal'? Is his worldview a product of poetic imagination or a product of metaphysical speculation, or is it based on hard science? According to him whichever path we follow, literature, metaphysics or science, we shall arrive at the same conclusion: 'This world is a human world, and the Universe is in harmony with Man, the eternal.'[11]

Tagore refers to the Universe as one systematic whole and to all branches of knowledge as being unified. In the 1930s the western scientific community was also strongly arguing in favour of the unification of science. The Vienna Circle, for instance, was scouting this view. Tagore went a step further and argued for the unification of all our investigations into truth. In support of his position he forwarded a monistic ontology and an epistemology based on the presence of human personality. For him all knowledge is human knowledge; he rejects the possibility of 'a view from nowhere.' We shall now present a more detailed account of Tagore's notion of the human and his notion of Man with an aim to facilitate a better understanding of the Tagore-Einstein conversation.

In the opening chapter of *The Religion of Man* Tagore endorses a notion of creation that is evolutionary in nature. His understanding of evolution is unique in many ways. He speaks of light as the radiant energy of creation. First came the planets, then life was brought into the arena of the creation-drama. In the next stage unicellular life-forms evolved into complex units of coordinated interrelationships. After this man appeared and turned the course of evolution. With this a transition takes place from the spirit of life to a spirit of Mind. The element of mind was all along present in nature, but its presence was made evident with the advent of man. While referring to mind at this stage Tagore spells it with a capital 'M.' For him Spirit, Spirit of Man, Spirit of Mind are synonyms of Mind. Man's nature is essentially spiritual. A new regime was introduced into life's kingdom with the emergence of man's consciousness of his own creative personality. Since personality is a key concept in Tagore's conceptual scheme, we need to go deeper into this concept.

With the emergence of mind in the scheme of things a distinct dualism occurs within nature – the previous life-world was guided exclusively by the laws of nature, but with the appearance of man freedom is introduced into the existing scheme. A new antagonism seems to have taken place in the shape of hostility between determinism and freedom. The uniqueness of Tagore's position is that he accepts the presence of binaries, but he never juxtaposes them as exclusive disjunctions – life/death, love/separation, day/night are examples of difference. These differences do not represent hyper-separation. Tagore calls the relation of mutual exclusion 'a dualism of hostility,' which he contrasts with 'a dualism of relationship.'[12]

In the initial stage of man's evolution he is overwhelmed by his power of freedom. His relationship with nature and its laws is denied or looked down upon; freedom and power of intellection are projected as the defining characteristics of man. In the initial stage man lays an excessive weightage on separation, often forgetting that human beings are jointly situated in the deterministic world of nature and the moral world of humanity, the world of freedom.[13] Tagore's continuity-thesis in the domain of evolution is unmistakable; man remains connected to the governance of nature while possessing the wings of freedom.

During the first stage of man's emergence the emphasis is on man's separateness from nature. With the help of science man tries to transform the hostile atmosphere of nature into a conducive habitat. At this juncture science, not spiritualism, plays the leading role in establishing man's freedom. Freedom at this stage is essentially 'freedom from. . . .' Our spirit seeks 'freedom from the slavery of things.'[14] At this stage of evolution the relationship between nature and culture appears to represent a hostile dualism. Science undoubtedly plays a liberating and empowering role. No other discipline can replace science in this endeavour. Referring to science Tagore says,

> She is working to give into man's hand Nature's magic wand of power; . . . the day will come when some of the great powers of nature will be at the beck and call of every individual, and at least the prime necessaries of life will be supplied to all with very little care and cost.[15]

Tagore had great faith in science; he also had great expectations. The enigma lies here. How does Tagore propose to accommodate his faith in science with his human conception of the universe? Faith in science is conventionally coupled with an acceptance of the autonomy of science. The enigma can only be addressed by focusing on Tagore's remarks on the relation of the world to science. The journey of evolution does not end with man's achievement of 'freedom from . . .' Tagore in his conversation with Einstein refers to science as, 'the impersonal human world of truths.' This may sound puzzling. How can a world be human and impersonal at the same time? The laws of nature are impersonal, they are not created by man. Tagore reminds us that in this world man is not the only factor to consider: there are other factors too, for instance natural laws. These laws are not alien to man. As a part of nature man is integrally related to nature and to its laws. In his *Sadhana* lectures the poet writes,

> We know that law cannot be set aside, and in this knowledge we have become strong. For this law is not something apart from us; it is our own. The universal power which is manifested in the universal law is one with our own power.[16]

In one sense we can call the whole world an extension of ourselves. The commonality between the world and me is one ground for referring to the world as a human world. Tagore uses the expression 'world-self.'[17] These are ontological grounds for considering the world to be a human world. The poet is referring to a material organic whole to which man also belongs. There are epistemic grounds too for upholding the notion of 'the impersonal human world of truth.'

Science leads us to an understanding of 'the universal in its physical aspect.'[18] In *The Religion of Man* Tagore speaks of a fictitious tiger that wants to wash off his black stripes with soap. The tiger in the story gains a distinct individuality not shared by other tigers. Science is not concerned with individuality, nor with such idiosyncrasies. Science is interested in the generic features of a tiger. Tagore says, 'A tiger must be like every other tiger in order that it may have its place in a book of Science; . . . We *know* a thing because it belongs to a class; . . .'[19] The scientist also belongs to a class. The class of scientists carry out their scientific investigation as a community. Unlike artists who travel alone, the scientists seek constant acceptance from their community. The artist's creativity does not come through generalizations. Tagore says, 'Science is a passenger in a railway train of generalization; there reasoning minds from all directions come to make their journey together in a similar conveyance.'[20] It is possible for all investigators to follow the same path because science is impersonal. 'The rational mind of the universal Man' is represented in science.[21] Science is impersonal in the sense that our scientific mind can be detached from our heart, our emotions. Being impersonal, science can move in the world of thoughts, in the world of the intellect, which is also impersonal. The impersonal intellect seeks to unfold the mysteries of the impersonal world. The scientist aims at being detached and unbiased in his investigation of the world.[22] If this be the case, then there seems to be no difference between Einstein and Tagore on this account: both seem to hold on to an objectivist point of view – objective meaning public. The term 'objective' has been used in many senses. The following are some of the commonly used senses:[23]

1 Objective means detached
2 Objective means public
3 Objective means existing independently
4 Objective means existing, the way things are

Numbers (2) and (3) above will unhesitatingly be accepted by Tagore. If by detached is meant unbiased or impersonal, then (1) will also be accepted by Tagore. For him (4) will be somewhat problematic. Let us now examine why this is so.

Tagore is in agreement with Einstein and other mainstream scientists that, 'Mathematics as the logic of number and dimension is the basis of our scientific knowledge.'[24] Tagore keeps shifting back and forth between the domains of metaphysics and epistemology. At the level of metaphysics he speaks of the infinite spirit assuming finitude. All the facts of this world are related to this central creative power. The infinite being expresses himself through the medium of finitude, the world. The infinite being is circumscribed by the self-imposed limitation of the medium of expression. The self-imposed limitations come in the form of the laws of space and time, of form and of movement. Tagore says, 'This law is Reason. . . .'[25] So the natural laws are also the rational laws. Science is concerned with knowing the laws of the finite universe. This is the direction in which science keeps moving. In principle science is capable of gaining an exhaustive knowledge of the laws of nature. Science also claims to achieve objectivity in the sense of (4) above – 'Objective means existing, the way things are.' No doubt this is the way the finite world is, but the finite world is only the medium through which the infinite is expressed. Science, however, is not willing to accept the paradox of the infinite assuming finitude. Tagore insists the paradox cannot be wished away.

According to Einstein the scientist is possessed by the law of causation. Tagore also upholds a similar view,

> In the scientific idea of the world there is no gap in the universal law of causality. . . . This is a generalization which has been made possible by a quality of logic which is possessed by the human mind.[26]

Along with the deterministic scientific idea of the world man also has the first-hand idea of freedom and he constantly struggles to expand his freedom. The co-presence of freedom and determinism has its roots in the paradox mentioned above – the paradoxical togetherness of the finite and the infinite. Notice that the universal law of causality is a generalization that is based on a unique human logic, which means we can conceive of other systems of logic as well. Tagore accepts the possibility of other fields of logic for other types of minds.[27] Not only logic, Tagore entertains the possibility of alternative space-time configurations as well. There could be a temporal world without space. Comprehension in such a world will be comparable to the comprehension of music. There are creatures in this world that know things which humans cannot comprehend. This is because 'their time is not synchronous with ours.'[28] Tagore does not say that there is no world out there, but the limits of my knowledge are the limits of my world (finite world). We do not receive a more or less accurate mirror image of the world.[29] Tagore says, 'The world, while I am perceiving it, is being incessantly created for myself in time and space.'[30]

Einstein was willing to accept that Beauty was not independent of man but he was not willing to accept Truth was not independent of man. It appears that Einstein took Tagore to believe if there is no human mind then there is no world either. That was not Tagore's contention. He held, if there is no human mind, then there is no humanly known world. He writes, 'I do not say that there may not be a truth which is beyond our utmost capacity of knowledge but for us it is as good as naught.'[31] The same thesis is reiterated in his *The Religion of Man*. There he argues that even if there were different worlds, it could not be proved.[32] Proof belongs to the domain of epistemology; in reality there may be other possible worlds not known to us. Tagore says, 'I do not imply that the final nature of the world depends upon the comprehension of the individual person.'[33] The final nature of the world depends on the Supreme Person, the infinite who has created finitude.

With the progress of science the wholeness of the world, its unity, is gradually becoming clear to us. The unity is known not merely through the extension of information. Tagore refers to the unity as a mystery. Unity is not confined to the finite, the unity of the infinite and the finite, the determined and the free, must also be comprehended. Progress is not limited to the domain of science. Scientific progress is the first step in the human journey of progress. By progress Tagore means that there is an ideal perfection to be reached, a point where all contradictions meet in the universal. The goal to be achieved through progress is not a theoretical construct, it is a fact. Tagore says, 'the great Spiritual Man is a fact to which we belong, . . . Man's movement is always progressively toward an objective reality.'[34] From this we could conclude that Tagore corroborates the fourth characterization of objectivity given above: 'objective means what there is – the really real.' However, we must add a caveat here, the Supreme Person according to Tagore is a creator who is constantly creating new forms and man in turn incessantly partakes in acts of creativity. This gives birth to a universal dynamism. The sojourn to progress does not tend at one absolute point of realization. Through evolution we are progressing towards the central truth, which is dynamic in nature.

The final stage of progress is reached not merely by extending the limits of knowledge and power, an extension of love and enjoyment through creative imagination is equally important. Tagore holds that imagination is the most distinctively human faculty.[35] Through imagination man is capable of transcending the deterministic world of causation and the universal reason of science too. Since imagination attempts at capturing the infinite, it cannot do so by deploying absolutely finite forms, forms must keep suitably changing.[36] Beyond science the path of progress towards truth follows a radically different path. Here according to Tagore,

> The guidance of reason constantly varies its course, in its perpetual process of adjustment with unforeseen circumstances; its scope is ever being widened by contact with new data.[37]

The adjustment of reason to data is an unending process since new data keep endlessly emerging.

Beyond science there is scope for a further qualitative growth of freedom. The scientist's detached position is important, the universal can be realized in the finite world only through it. At this final stage of evolution freedom shifts its focus from separation to unity as a result hostile dualisms get transformed to relational dualisms. This latter stage of freedom marks man's birth in the spiritual world. The spiritual world is not a transcendental world beyond the world of science, it merely marks a reorientation of man to the world of science. Science only represents the reasoning mind of the universal Mind. The universal Man is more than the rational mind. Beyond science the universal Man or Supreme Person has to be realized through imagination, love and wisdom.

Man now realizes that the world viewed in a detached manner is not as energizing and power-giving as the world viewed through relationships. The more energizing relations are not the original relationships handed down by nature. New relationships need to be repeatedly forged through man's creative imagination for the sake of greater harmony, happiness and fulfilment. Man is not content with the world as it is given to him, he constantly tries to rearrange things. In *The Religion of Man* Tagore writes,

> In order to live efficiently man must know facts and their laws. In order to be happy he must establish harmonious relationship with all things with which he has dealings. Our creation is the modification of relationship.[38]

Thus, through evolution man gradually passes from a world that 'is' to a world that 'ought to be.'

The 'ought to be world' is created by man's personality. Consciousness of personality begins with the feeling of separateness and ends with the feeling of unity with all. Tagore holds that the infinite dwells in the ideal of unity. At a deeper level all things of this world are related. Unity exists in the world, the unity of relationships does not exist in space. Unity resides in the spirit of things and it can only be realized in one's own spirit.[39] There is another kind of external physical unity constructed by science. Vehicles of communication for instance have been breaking geographical barriers and bringing individuals closer. Tagore laments that the inner forces of unity that bring men together are lagging behind.[40]

Personality creates its own inner nature as well as its outer surroundings. Personality has a unity with oneself as well as a sense of unity with others. Through the expression of his personality man establishes himself as a creator and identifies himself with the creator of the universe. Thus arises a consciousness of unity of purpose as well as a unity with the world. In the first phase of his freedom man's identity is expressed through the notion of self. The assertion of freedom consists in the separation of the self from nature. At this stage nature is man's 'other.' In the second stage of man's freedom he passes from self-consciousness to soul-consciousness. In the beginning personality has the sense of separateness from all, then it gradually proceeds to the feeling of relatedness to all. Soul-consciousness gives man the realization that in the heart of his world he has met the Immortal Person – he has direct communication of the person with the Person.[41] The finite person communicates with the infinite Person within his own being. This implies that the individual in the final stage of evolution is both finite and infinite, both person and Person. Is this not a contradiction?

Admittedly there is a contradiction, but then this is the nature of reality: it is both finite and infinite, simultaneously and not sequentially. To project either infinity or finitude as the ultimate reality is a fallacy. The finite and the infinite need to be harmonized into a single unity. New unities keep emerging. This is an ongoing process. Tagore is often referred to as a process philosopher. The eternal Spirit does not transcend the process; it is part and parcel of the process. The world is not a construct created once and for all. Tagore always refers to the world as a creation. The world has a grammar and a logic, which it surpasses in its dynamism. We humans as creators incessantly modify our relationship with the changing world.

Tagore says, 'But truth is beyond logic; it is the everlasting miracle; it is static and dynamic at the same time; it is ideal and real; it is finite and infinite.'[42] On the one hand, there is the truth of the finite world arrived at through science, and on the other hand, there is the truth of the spiritual world which unites the finite with the infinite. The former knows truth in its parts and the latter knows truth in its whole.[43] These are two separate systems. Their investigative apparatus are different though they are speaking of the same human world. In a scientific discourse contradictions are considered to be unintelligible. How then does one arrive at the conclusion that reality is both finite and infinite? Tagore says, 'it cannot be explained, it has to be realized.'[44] It is a kind of self-evident truth. Since the infinite cannot be measured, it cannot be defined either. This knowledge comes to us through immediate perception; it is not attainable through reason.[45] Such knowledge comes to us immediately with a conviction.[46] No further justification is needed. This, however, does not make reason superfluous. In the finite world of 'freedom from . . .' reason is the supreme guide. Yet man must not be identified with reason. Man is identified by personality. Reason is merely a useful tool.

Tagore speaks of three dimensions of man's identity: the physical, the mental, and the spiritual. Man's body is a marvellous machine, which is totally guided by the laws of nature, man's mind is an organ of efficiency guided by reason. Over and above being a mind and a body man is also a soul. The soul is the creative power in man; however, it remains dormant until it is recognized. The soul is revealed when we know that the One in us is the One in all. The soul is a central creative power. While my mind and body are exclusively mine, the soul is not exclusively mine, it is intimately mine. While realizing the unity with the One the individual does not sever her relationship with the physical and mental world of the many. The soul through its freedom creates new relationships with the world. Imagination plays a major role in this context. A balanced account of life must take both aspects into account, the physical as well as the spiritual. A harmony between the two can only be established through sufficient detachment of mind. The unchangeable scientific laws of nature cannot be wished away by the free spirit, nor can the spirit accept unquestioned subjection to these laws. In order to establish self-rule of the spirit the new relationships created by the spirit need to be harmonized with the scientific laws of nature.

There are innumerable personalities; each is a source of creativity. How can they be reconciled into one unified frame? If there is an iota of truth in that which is created, it must be universal, it could not remain personal. Truth can never belong to the particular country which produces it.[47] No doubt we are individuals. As individuals we keep changing, we keep coming close to each other; we mix and mingle till we realize our unity with the universe. 'This consciousness of oneness beyond all boundaries is the consciousness of soul.'[48] The *Isha Upanishad* speaks of knowing the soul beyond boundaries.

Consciousness keeps expanding; it keeps on surpassing boundaries till it realizes its unity with the all-feeling being. This ultimate stage of consciousness is love, the summit of consciousness.[49]

The personal and the impersonal both unite in love. Love leads to harmony between opposites – the finite and the infinite, the separate and the united, the near and the far, the one and the many are all united through love. Tagore says, 'In love all the contradictions of existence merge themselves and are lost.'[50] Love does not shy away from limitations, it transcends them. Love assures me that the world is united with my soul. Man's highest truth does not reside in separation; the highest truth lies in co-operation and love. Love is an emotion. It is not reason that leads us to the summit of consciousness; it is the emotion of love. The revelation of reality is immediate, it needs no medium. Emotion and imagination create the background of our mind against which reality is manifested.

Self-interest causes an obstruction to the vision of unity. Tagore felt contemporary European science was being used in the service of utilitarian goals and self-interest and, therefore, its vision of unity was being obfuscated – science was proceeding with a tunnel vision. Utilitarian goals are achieved through desire-fulfilment. Desire-fulfilment leads to further separation – desires are an impediment to unity. Desires keep multiplying and they lead to greater and greater consumption. Greater consumption means greater exploitation of nature – greater separation between man and nature. Desire obstructs unity. Tagore is not propounding a Stoic ideal bereft of desires and utilitarian goals; the underlying tenor should be one of moderation. He says, 'Of course man is useful to man.'[51] Beyond his use-value he also has a surplus out of which he creates – he is an artist and a lover. These are of greater value than his use-value. Science is welcome as long as it serves the purpose of our needs, but when it is used for the purpose of greed-satisfaction, we are wooed away from the path of the spirit, the path of unity. When science leads us on the path of separation, we are made to believe separation is absolute, which is not true. We must move beyond separation to unity. By contrast, literature and art can broaden our horizon – they can take us beyond the narrow boundaries of separation and finitude. Art is capable of giving us an unobstructed vision of unity. Tagore believes the scientist can also broaden his horizon by going deeper and deeper into his subject to a point where science itself turns into metaphysics.[52]

After this elaborate exegesis of Tagore's position on science, art and ontology we are in a more informed position to judge the Einstein-Tagore debate. We began by saying, Tagore was speaking from the position of Upanishadic idealism and Einstein from the position of scientific realism. Before we re-enter the debate it would be helpful to very briefly examine Einstein's appreciation of Spinoza's philosophy. Einstein was deeply interested in Spinoza. This is evident from his letters and conversations. It is well known that Spinoza was a monist and a pantheist, he did not believe in a personal god, neither did Tagore or Einstein believe in a personal god.

Einstein was not an atheist, at the same time he was not sure whether he was a pantheist or not. He believed in Spinoza's god as far as god revealed himself in the lawful harmony of all that exists. For Tagore the Supreme Spirit also reveals himself in the lawful harmony of all that exists. In other words lawful existence is a manifestation of god. Tagore's god has a dual manifestation as law guided finitude and as the free creative infinitude. Human beings are also guided by the lawful harmony of the world and are free to create too. A full understanding of finitude and infinitude necessitates an understanding of personality. Personality explains the presence of freedom in the scientist and in the artist. For Tagore the world is infused by the presence of spirit either in a manifest form or as 'a yet to be manifest' potential. It was on this account that we referred to Tagore as an idealist. In the history of philosophy Spinoza is known to be a determinist. When Einstein expresses his allegiance to Spinoza's position, it is his unrestricted determinism that singularly attracts him. According to Spinoza god's will is identical with the laws of nature. In Spinoza's system individual human beings are not free. They have no alternative to their actions.

On the issue of unrestricted determinism Tagore differs from Einstein. True, the world is determined by the laws of nature. Man's surplus and freedom, however, surpass the domain of determinism. Tagore's scheme acknowledges the presence of a vast area for self-expression of the spirit. When Einstein recognizes the revelation of a superior intelligence in nature, he comes very close to idealism. Einstein the philosopher comes close to Spinoza, Einstein the physicist does not. A parallel comparison could be made with Tagore – when Tagore deals with the broad outline of science, he comes very close to Einstein; when Tagore deals with religion, he moves worlds apart.

Notes

1 'Einstein and Tagore' (conversation), in Sisir Kumar Das (ed), *The English Writings of Rabindranath Tagore* (*EWRT*), vol. 3, New Delhi: Sahitya Akademi, 1996, p. 911.

2 Rabindranath Thakur, *Biswa Parichaya, in Rabindra-Rachanabali* (in Bengali), vol. 25, Visva-Bharati Edition, 1937.
3 Parimal Goswamy, 'Rabindranath O Bigyan', in Pulinbihari Sen (ed), *Rabindrayan* (in Bengali), vol. 2, Kolkata: Bak-Sahitya, 2007, p. 164.
4 Ibid., p. 167.
5 Rabindranath Tagore, '*Sadhana*', in *EWRT*, vol. 2, 1996, pp. 277–345.
6 '*Personality*', in *EWRT*, vol. 2, pp. 347–416.
7 'Preface' to *Sadhana*, in *EWRT*, vol. 2, p. 278.
8 *The Religion of Man*, in *EWRT*, vol. 3, 1931.
9 Rabindranath Tagore, 'My Memories of Einstein', in Sanjib Mukhopadhyay (ed), *Rabindranath-Einstein: Ek Amimamgsito Sanglap* (*Tagore-Einstein: An Unresolved Dialogue*) (Bilingual: English and Bengali), Kolkata: Sahitya Samsad, 2014, p. 339. Also reprinted here as Appendix A.
10 'Preface' to *Sadhana*, in *EWRT*, vol. 2, p. 278.
11 Goswamy, 'Rabindranath O Bigyan', in *Rabindrayan*, p. 157 (translation mine).
12 *Personality*, in *EWRT*, vol. 2, p. 377.
13 Ibid., p. 378.
14 Ibid., p. 382.
15 Ibid.
16 *Sadhana*, in *EWRT*, vol. 2, p. 304.
17 Ibid.
18 Ibid.
19 *The Religion of Man*, in *EWRT,* vol. 3, pp. 141–2.
20 'The Meaning of Art', in *EWRT*, vol. 3, p. 587.
21 *The Religion of Man*, in *EWRT*, vol. 3, p. 91.
22 'Thoughts from Tagore', in *EWRT*, vol. 3, p. 77.
23 The classification is based on Elizabeth Lloyd's article 'Objectivity and the Double Standard for Feminist Epistemologies', *Synthese*, 1995, 104(3) (September), p. 375.
24 'Thoughts from Tagore', in *EWRT*, vol. 3. p. 77.
25 *Personality*, in *EWRT*, vol. 2, p. 368.
26 *The Religion of Man*, in *EWRT*, vol. 3, p. 90.
27 Tagore, 'My Memories of Einstein', in *Rabindranath-Einstein: Ek Amimamgsito Sanglap*, p. 341.
28 *Personality*, in *EWRT*, vol. 2, p. 364.
29 Ibid., p. 365.
30 Ibid.
31 Tagore, 'My Memories of Einstein', in *Rabindranath-Einstein: Ek Amimamgsito Sanglap*, p. 341.
32 *The Religion of Man*, in *EWRT*, vol. 3, p. 91.
33 Ibid.
34 Tagore, 'My Memories of Einstein', in *Rabindranath-Einstein: Ek Amimamgsito Sanglap*, p. 341.
35 *The Religion of Man*, in *EWRT*, vol. 3, p. 104.
36 *Personality*, in *EWRT*, vol. 2, p. 370.
37 'Thoughts from Tagore', in *EWRT*, vol. 3, p. 80.
38 *The Religion of Man*, in *EWRT*, vol. 3, p. 139.
39 'Thoughts from Tagore', in *EWRT*, vol. 3, p. 81.
40 'The Union of Cultures', in *Lectures and Addresses*, *EWRT*, vol. 3, p. 435.

41 *Personality*, in *EWRT*, vol. 2, p. 374.
42 'Letters to a Friend', in *EWRT*, vol. 3, p. 248.
43 *Personality*, in *EWRT*, vol. 2, p. 364.
44 Ibid., p. 375.
45 'The Principle of Literature', in *EWRT*, vol. 3, p. 595.
46 Ibid., p. 691.
47 'The Way to Unity', in *EWRT*, vol. 3, p. 465.
48 *Personality*, in *EWRT*, vol. 2, p. 373.
49 *Sadhana*, in *EWRT*, vol. 2, pp. 320–1.
50 Ibid., p. 324.
51 Ibid., p. 322.
52 Ibid., p. 320.

References

Goswamy, Parimal. 1961. Rabindranath O Bigyan. In Pulinbihari Sen (ed), *Rabindrayan* (in Bengali), vol. 2, Kolkata: Bak-Sahitya, 2007.

Lloyd, Elizabeth. 1995. Objectivity and the Double Standard for Feminist Epistemologies. *Synthese*, 104(3) (September): 375.

Tagore, Rabindranath. 1931. My Memories of Einstein. In Sanjib Mukhopadhyay (ed), *Rabindranath-Einstein: Ek Amimamgsito Sanglap* (*Tagore-Einstein: An Unresolved Dialogue*), (Bilingual: English and Bengali), Kolkata: Sahitya Samsad, 2014.

———. 1931. *The Religion of Man*. In Sisir Kumar Das (ed), *The English Writings of Tagore*, vol. 3, New Delhi: Sahitya Akademi, 1996.

———. 1930. Einstein and Tagore' (conversation). In Sisir Kumar Das (ed), *The English Writings of Tagore*, vol. 3, New Delhi: Sahitya Akademi, 1996.

———. 1928. Letters to a Friend. In (Sisir Kumar Das), *The English Writings of Tagore*, vol. 3, New Delhi: Sahitya Akademi, 1996.

———. 1927. The Principle of Literature. In Sisir Kumar Das (ed), *The English Writings of Tagore*, vol. 3, New Delhi: Sahitya Akademi, 1996.

———. 1926. The Meaning of Art. In Sisir Kumar Das (ed), *The English Writings of Tagore*, vol. 3, New Delhi: Sahitya Akademi, 1996.

———. 1923. The Way to Unity. In Sisir Kumar Das (ed), *The English Writings of Tagore*, vol. 3, New Delhi:Sahitya Akademi, 1996.

———. 1921. Thoughts from Tagore. In Sisir Kumar Das (ed), *The English Writings of Tagore*, vol. 3, New Delhi Sahitya Akademi, 1996.

———. 1921. The Union of Cultures. *Lectures and Addresses*. In Sisir Kumar Das (ed), *The English Writings of Tagore*, vol. 3, New Delhi: Sahitya Akademi, 1996.

———. 1917. *Personality*. In Sisir Kumar Das (ed), *The English Writings of Tagore*, vol. 2, New Delhi: Sahitya Akademi, 1996.

———. 1913. *Sadhana*. In Sisir Kumar Das (ed), *The English Writings of Tagore*, vol. 2, New Delhi: Sahitya Akademi, 1996.

Thakur, Rabindranath. 1937. *Biswa Parichaya*. In *Rabindra-Rachanabali* (in Bengali), vol. 25, Kolkata: Visva-Bharati Edition, 1970.

6

POSITING THE HUMAN IN THE HUMAN SCIENCES

Sourin Bhattacharya

The celebrated Tagore–Einstein conversation in Kaputh, Germany in July 1930[1] has rightly drawn much academic attention. There are, however, many controversies around the issues that were raised or were supposed to be raised in the conversation.[2] It is not exactly clear if what the two great men had talked about was very correctly recorded and then subsequently adequately edited before being published. There are suggestions that Einstein was perhaps not happy with the way the matter was brought to public attention. Besides, there was also a problem of language difficulty in that Einstein was not very comfortable in English and Tagore would not speak German. Was there an interpreter? Perhaps not.

At present I am not directly concerned with the debate itself. I wish to focus on one central issue in the conversations and deal with its implications in the human sciences. The central issue is how to posit man. Einstein was a thorough realist in these discussions. He believed in the existence of an external world independent of any human gaze. There is a world lying out there. Sciences take upon themselves the task of unfolding the mysteries of its working. Scientific theories are but designed to give explanations or descriptions of how nature is organized and how its constituting elements are structured in their interrelation. In a kind of realist view mediation of the human mind is of no consequence in the constitution of reality. The study itself is not precluded though from being influenced by human intervention. The nature of reality and its comprehension are two separate concerns. Tagore, it seems, was preoccupied with the nature question rather than the question of comprehension. That is reasonable enough. To comprehend is the primary responsibility of a scientist *per se*. Tagore was not obviously arrogating that task for himself. He knew the way he realized himself as but an element in all that surrounds him in the phenomenological sense. He is not merely an outsider. He is passionately in it. In a letter to Ramendrasundar Trivedi[3] he writes in an admonishing tone that what they would consider merely his poetic stance was not anything exterior to him, it was his real nature. His poetry emanates from that nature. He writes poetry because he is what he is. The occasion was that Ramendrsundar had suggested while editing the letters collected in *chinnapatra* that certain portions of a particular letter should better be left out for it contained an unacceptable amount of poetic ecstasy. Tagore's point was that the ecstasy was in his nature; it was not something artificially imposed upon himself. His poetry was *his* poetry. The ecstasy in question was an expression of what he felt about the trees and the seas and all that came here on this earth before his appearance. As he came later he was humble enough to recognize that he would be a possible intruder to the predecessors, but as a latecomer he knew them all. This is how Tagore felt himself to be an insider to everything here. So nothing is out there for him. It is all *in* him. This is clearly not a scientist's approach as science is normally practiced. That perhaps was the communication barrier between Tagore and Einstein. A student of Tagore music and poetry would be reminded of those two birds; one of which was caged and the other was free in nature. The two could never share their language of anguish though both had made good friends for a while.[4] Neither of them understands each other, nor do they make themselves understood. This miscommunication might have led to misgivings on either side, maybe more for Einstein than for Tagore.

Tagore in this debate introduced a very important concept of universal man. This was a central theme in his Hibbert lectures, which were co-temporal with this German visit, which was the occasion for his meeting with Einstein. In fact when the lectures were published in book form as *Religion of Man* the conversations appeared in the appendix.[5] It may be supposed that that was the time when Tagore was seized of this idea. As a matter of fact the idea of universal man was a basic element in Tagore's thought. He talks about a kind of evolution of this idea in his mind. It was sort of organic growth with him. At the beginning he does not use the term 'universal man.' He conceptualized a Man's universe with a capital M 'and we must acknowledge,' writes Tagore, 'that the evolution that Science talks of is that of Man's universe.'[6] The appearance of man in the universe is for Tagore a point of singularity as though, 'Light is the radiant energy of creation.' That begins it all. Then from unicellular life to multicellular life to ultimately man's life. True as a single instance it was a particular life but that particularity assumed lasting significance only in that it signified the appearance of man with that capital M.

That is the story of evolution of the world 'and this world that we perceive through our senses and mind and life's experience is profoundly one with ourselves.'[7] The profundity of this being one with ourselves is something that Tagore seemed to fail to communicate to Einstein. Understandably perhaps. The cultural background of the two noble minds was different and this perhaps caused difference in sensibilities and responses. Maybe we have to come back to this question later on. But for now I move on to consider some debatable issues that provide a background for positing this Tagore Man in the human sciences.

Sciences-war perspective

I have been following the celebrated Academy Debates reported by two reputed scientists, physicists as a matter of fact, Tony Rothman and George Sudarshan in *Doubt and Certainty*.[8] These are delightful reading for a non-specialist like me, who has some interest in scientific themes without having adequate technical competence to pursue them. I place much emphasis on such writings for the simple fact that without them people like me are bound to be left in a soup. The scientific community normally appears to be esoteric. They do not bother about the outsiders. Why should they? Even if the members of the lay public applaud some of their activities and understandings that would mean nothing to them unless what they say appears in a peer-reviewed journal or is acclaimed in any recognized academic conference. That is normal practice in science. Scientists cannot be faulted for a craving for recognition from one among their ranks. One thing may be plainly admitted. However strong may be the case for objective knowledge claim, the acceptability question remains problematic. Proper recognition is but one way to move towards that goal. The fierce debate between the scientists of rank that often erupts regarding the meaning and significance of proposed theories are a pointer to the presence of extra-objective elements ranging from scientific prejudices and predilections to even personal animosity perhaps. In addition there may be problems of state intervention, ideological positions, historical compulsions and national demands. Research programmes in science, especially in today's rigidly organized social structure, are less than 'pure' in most cases. Knowledge claims also need to be adjusted accordingly.

Frankly speaking, I allow myself to be led by Rothman and Sudarshan through the corridors of their celebrated academy in this bad faith that I may be privileged to watch the men of reputation assembled here washing some of their dirty scientific linen while thrashing out their theoretical differences. I confess I belong more to the doubt part of the title rather than the certainty part of it. Despite the clamour for experimental findings I feel our assertions need to be humbler. All this is a kind of our mental arrangement. Whatever we say about the world we live in, including the social and the human world, should always be taken to be related to the way we have so far ordered things in our mind. Take the well-known instances of 'the end of history' and 'the clash of civilizations' theories.[9] These authors have ordered national developments and international events in a manner that the collapse of Soviet socialism and the rise of Islamic extremism appeared to them in that light of triumph of liberal democracy. What had happened had happened in reality. We build up a narrative on the events claimed, taken or reported to have happened. The events are there. The narrative is ours. The events in themselves do not constitute the reality. The narrative we build up shapes the social reality. In what sense can we then say that the social reality lies out there? But we often take shelter under that pretension in order to employ the methods of natural sciences in matters dealt with in the human sciences. This explains my somewhat mischievous motive in following the Rothman-Sudarshan lead.

I enter the tenth debate. The subjects of the first nine debates are:

1 Is the universe describable?
2 Is nature unreasonably mathematical?
3 Is the world symmetrical?
4 Why do things happen?
5 Does time go forward?
6 Why is there left and right?
7 Is the universe weird?
8 Is there an answer?
9 How did we get here?

Now comes the tenth debate with the title 'What do you mean?' Like other debates this one also has a descriptive subtitle – metaphors, analogies, culture wars. The earlier debates were on subjects ranging from laws, principles, theories and models to causality, synchronicity and all that to quantum mechanics to theories of everything, general relativity and the standard model. These are admittedly well-known topics of the kind of discourse generally regarded as science. While pursuing these topics we now come to the subjects of metaphors, analogies and culture. These are on the other hand some of the topics usually dealt with in humanities and social sciences. I have been so long following the goings-on in Rothman-Sudarshan academy for this moment. The reality we live in is a totality. There can be no meaningful divide between the sciences and the rest, if the reality itself is our subject of knowledge or understanding. Maybe in different disciplines we approach the subject differently, but unless we suppose the reality to be comprehensively total it becomes by definition fragmented. The ontology of it gets epistemologically determined. These two levels need to be kept separated and contemplated one in isolation of the other. To contemplate the nature of reality is to try to shape something in one's mind, irrespective of how one proposes to know that reality. But things become immensely complicated if it turns out to be that the what question is at least partly shaped by the how question. This is precisely where a veritable war takes place between the sciences on one side and the humanities with its allies on the other. No wonder that the subject description of the tenth debate openly uses the phrase 'culture wars.' It was a regular war situation indeed. This has been even otherwise described as sciences war.

It all began with the publication of an article titled 'Transgressing the boundaries: Towards a Transformative Hermeneutics of Quantum Gravity' by Alan Sokal, a Physics professor at the New York University in *Social Text*, Science Wars issue of May 1996. The paper argued that quantum gravity is a social and linguistic construct. The paper eventually turned out to be a hoax. On the day of publication of this article Alan Sokal himself revealed in *Lingua Franca* that he had perpetrated it to sort of mock at the postmodern way of seeing things and writing them in a verbose deconstructionist style. One obvious target was Jacques Derrida.

There followed a fierce debate with 'Transgressing the boundaries' article as the reference point. Now several things may be noted here. One is that the author's own hoax claim should not be taken too seriously. Who knows that the claim itself is not a hoax again? So it needs to be independently judged. By whom? That is the crux of the problem. The whole scenario was made so combative. And that too not necessarily by just the publication of this article. Alan Sokal used such phrases as 'barbarian hordes of lit crit.' The warring groups were the sciences men and the humanities men. This belligerent atmosphere is much older than the Sokal affair. I am coming to that. But given the belligerence reasonable and credible judgement is hardly possible. If it is by the sciences men that will either not carry much weight with the humanities men or only carry the weight of an authority and if it is by the latter then the accusation of non-understanding will be flung at them. A huge communication gap.

The next thing I should like to draw attention to is the emergence of postmodernist scholarship. There are strong reactions against some of the ideas of this style of thinking and writing. Postmodernism is after all no homogeneous thing. It has many shades and many nuances with no neatly organized coherence in its critical thrust. But still there is some marker characterizing postmodernist features. Writers in this style often take recourse to exaggerations. And the style of writing is somewhat unfamiliar in that it may seem jargon ridden and heavily loaded with use of neologisms. But to be fair one thing has to be admitted. If new ideas are to be introduced then they may bring with them new ways of expressing things as well. Think of poetry. New generation of genuine poetry brings with it new poetic diction. It also happens that that new idiom takes time to go down with the readers. This happens not just in poetry. The scenario is roughly the same in painting, theatre, films and music and perhaps in almost any other meaningful creative activity of the human mind. Think of even hard sciences. New paradigmatic breakthroughs often need newer analytical techniques, like new kind of geometry and mathematics. This is a very common experience. New techniques of today may get older and inadequate at a later date with the emergence of new problematic and new approaches to handle them. That also generates tensions between different generations. In economic theory Paul Samuelson once used to be looked upon as a hard mathematical economist and was known for being single handedly responsible for importing concepts and methods of mechanics and classical theoretical physics of the time. This was in the 1930s and 1940s. In the 1870s when economic theory was taking a turn towards marginalist ideas William Stanley Jevons was arguing for the introduction of classical calculus. Coming forward in the 1960s and 1970s when some core areas of economic theory was turning to axiomatization through the works of Kenneth Arrow and Gerard Debreau and the likes of them, Samuelson, already old-fashioned perhaps, became critical of this high profile mathematics. Now what I want to mean is that such critical and counter critical postures are nothing new or strange in learned academic circles. What generates unwanted bitterness is perhaps due to uncritical intolerance. I am not sure if something like that is happening here. Change of language, style and idiom in many cases is related to change of theme and substantial issues, connected with paradigmatic shifts. Superficial gimmicks are possible here and there, but they normally do not determine the course of development of strands of thought. That is what should hold our attention.

Two cultures theory

The debate following the Sokal affair is bound to bring to mind another earlier debate of a similar nature. The scientist-writer Charles Pierce Snow in his Rede Lecture of 7 May 1959 in Cambridge[10] introduced the idea that western mind is sharply divided between these two poles – the sciences and the humanities. The poles were so far apart that one would not see the other's face. Dialogue was hardly possible. Sciences gradually became technical and laboratory dependent and thus practically inaccessible to outsiders. Humanities, as it was written in plain prose, appeared to be within everybody's easy reach. This was of course an undue simplification having its origin in an uncritical approach to things seemingly free from technicalities. There followed a serious lack of communication that led to an atmosphere of mutual disbelief. But that was not all.

There were all the trappings of a power game as well, it seems. This works in at least two ways. One was evident in Snow's own account. Snow identified three critical problems in his times. These were nuclear weapons, over population and the gap between rich and poor nations. In view of these pressing problems the literary intellectuals appeared to Snow as 'natural luddites.' His complaint was that they failed to see that solutions might come from 'natural scientists.' Snow felt that the scientists held 'the future in their bones' and the humanities men would not recognize that. This is a position that can be argued to be an instance of almost a teleological faith on the power of science, in fact technology in the name of science. The power of science in this sense is so great that such a belief has filtered down to the common masses. Come any problem, there will soon be found some innovative measure. This is the kind of response often shown to the prophets of despair, who cry foul in anything from carbon emission to seawater salinity to global warming. No need to panic, the optimists would say. Science would sure show way to some technological way out.

Another source of a sense of power is the epistemological position as regards the knowledge status of science discourse. Science gives rise to genuine knowledge about the functioning of the real phenomena. Its genuineness is authenticated by experimental exercises, which the other human discourses have no access to. This amenability to experimental replication leads further to a claim of objective knowledge. Subjective claims are unreliable and ultimately depend upon forces of authority or appeal to aesthetic satisfaction, emotional comfort and all that. Only science can give us objective knowledge and we can negotiate our life processes better when equipped with that kind of knowledge. Now this question of negotiating the life processes, if it is to be taken seriously then that would probably diminish to some extent the claim of higher knowledge status for science statements. Life processes are much more comprehensive than the technological reach. In the two cultures thesis Snow was as emphatic as to say that 'If scientists have the future in their bones, then the traditional culture responds by wishing the future did not exist.' The humanities men are thus painted pessimists. The controversy grew bitter. F R Leavis on the other side called Snow 'a public relations man.'[11] Leavis's main intervention was in the course of The Richmond Lecture of 1962. The debate flared up after this lecture. The battle lines between the sciences and the humanities were drawn so sharply that in 2008 *The Times Literary Supplement* included C P Snow's *The Two Cultures and the Scientific Revolution* in its list of the 100 books that most influenced western public discourse since the Second World War.

The central methodological issue involved in the two cultures debate may be looked upon in a different perspective. There are suggestions from the science side that scientific statements can be expressed through 'is' sentences. And for that reason these sentences are statements about what the world is like. They say something about the state of the world. As though these statements are free of human valuation. This is the kernel of the objectivity claim advanced in sciences. This is also linked to the common feeling that ascribes a higher knowledge status to the science statements. In contrast to this positivist position German social sciences advocate the method of interpretation,[12] in which the human position takes precedence. If positivism is pretending that the object of knowledge is outside the subject the method of interpretation or understanding assigns a clear position to the human knower. The German word *verstehen* underlines this understanding aspect. This is about deriving meaning. In this the objective existence of things are informed by what it means to the gazer, maybe also along with to those who matter in this gaze. From this point of view the two cultures debate may appear to be a quarrel of positivism vs. interpretivism. The quarrel surfaces and resurfaces in different times in different forms without a hint of being settled either way. Larry Laudan writes:

> The displacement of the idea that facts and evidence matter by the idea that everything boils down to subjective interests and perspectives is – second only to American political campaigns – the most prominent and pernicious manifestation of anti-intellectualism in our time.[13]

Positivism is committed to facts and evidence. Hard facts indeed. On this basis a typical positivist claim is that things can be said clearly only on such bases. No other claim should have a place in scientific discourse. In response to this even Werner Heisenberg would have said if we wish to say things only on the basis of hard facts then possibly we would end up saying only trivial tautologies. To mean to say meaningful things we must be prepared for statements not exactly clear, maybe vague even.[14]

Relational structure and reality

Esoteric science is not for all. A majority is left out. Society suffers and it becomes that much poorer. The divide that is being talked about in the sciences war or the two cultures theory is at least partly due to the technical language in which modern science is written. There may be other reasons as well. Even when we have the good will on either side the technical difficulties may prove a real barrier. But for reasons cultural, if not for anything else, we need science. Tagore perhaps expressed some such thing when he said:

> The universe kept its own microcosmic self hidden, made its macrocosm suitably small or had taken it away behind a curtain. The universe presented itself in a form comprehensible to man within the bounds of his normal powers. But man is not simple. He is the only one to question his own perception, and raise objections. He is happy to vanquish his opponent. Man in his quest to transcend the boundaries of his human capacity has made the distant near, has made the invisible directly observable, has given verbal expression to the inexpressible. Man has entered the deep recesses of the mysteries of nature to go beyond the phenomenal world. The efforts that have made this possible is not accessible to many for want of ability and opportunity. And those who are deprived of this result will be thrown into the backyards of the modern era.
>
> In a big forest the dried leaves drop naturally down. That makes the soil fertile. In the land of cultivation of science the small things of knowledge are regularly scattered. That gives a vibrant mind a scientific turn. As we lack it our mind remains unscientific. This is not just a poverty of knowledge. We are dwarfed in our world of action.[15]

This vision of the poet that the tiny small things are hidden from human sight and the very large shows itself up in a form that man can grasp in his power and sensibility, that he is just not awed by an oceanic feeling, are the two limits within which Tagore posits the man who makes his science. Here one is reminded of Einstein's position that Lord God is sophisticated but not malicious.[16] He who makes science is Tagore's universal man, who is endowed with consciousness. For Tagore the nature of reality is what is filtered through to this conscious mind constituted as a unified whole. Proceeding along these lines Tagore develops a concept of creative personality,[17] which may go a long way to bring the 'two cultures' divide to a harmonious unity. But that is a different exercise. For the present we take up the image of leaves dropping down in a big forest that makes the soil fertile. This has an implication for simplicity.

Stephen Hawking in his seven lecture series entitled *The Theory of Everything*[18] gradually moves towards a position of simplicity. Hawking begins with that it is difficult to achieve a unified theory at one go. So the strategy should be to move through partial theories. The ultimate goal is 'to find a complete, consistent unified theory that would include all these partial theories as approximations. The quest for such a theory is known as "the unification of physics".'[19] Hawking alludes to Einstein's unsuccessful search for such a unified theory. He feels that time was not ripe then. Knowledge was still inadequate. Not much was known about the nuclear forces. The prospect is better in our time now. Hawking then travels through the current developments, especially in string theory, and following the question of dimensions arrives at an anthropic principle like this: 'Two space dimensions do not seem to be enough to allow for the development of complicated beings like us.'[20] This position is important for the argument we are trying to build up. Dimensions are essential components of the nature of physical reality in that it crops up in our space-time description, and also of its comprehension. But they are as essential for the existence of life as we know it. Here it is to be noted carefully that although Hawking calls this principle anthropic, it is not anthropocentric in any sense. It concerns not simply man's life; what is concerned is life as such as is known or as it has evolved up to now. Stephen Hawking finally takes a position of 'cautious optimism' about the prospect of unification.

If, however, unification is ever achieved 'that will be just a set of rules and equations.' 'What is it,' asks Hawking, 'that breathes fire into the equations' and that makes a describable universe? This is the why question, which is usually left out by the scientists as being not within their scope. We are reminded of the first *sloka* of *Kenoponishad* that runs like this:

> keneshitang patati preshitang manah kena pranah prathamah praiti yuktah |
> keneshitaang vacamimang vadanti cakshuh srotrang k u devo yunakti ||
>
> By whom missioned falls the mind shot to its mark? By whom yoked moves the first life-breath forward on its paths? By whom impelled is this word that men speak? What god set eye and ear to their workings?[21]

With this why question Hawking comes near the 'two cultures' crisis. That is now around the what and the why questions. One is for science, while the other is for philosophy. Now Hawking's optimistic conclusion is:

> However if we do discover a complete theory, it should in time be understandable in broad principle by everyone, not just a few scientists. Then we shall be able to take part in the discussion of why the universe exists. If we find the answer to that, it would be the ultimate triumph of human reason. For then we would know the mind of God.[22]

Unification and simplicity meet together. Sciences now become human sciences. Because the discourse that is generated at this moment of unification involves the human beings qua human beings and presumes a relational structure in the world of the living and the non-living. These are two different layers held in a common structure.

Conclusion

By way of conclusion we propose a conceptualization of reality through the notion of a relation space. Let R={Ri} be a set of n-ary relations defined over an infinite set of elements X= {xj}. While i is taken to range over an arbitrarily finite number of values, j is allowed to range from 1,2,3, to infinity. In other words we visualize a situation in which we can have at any time a finite number of relations defined over a part of an infinite number of elements. A discourse set D is defined to be D={X;R}. Our sense of reality is conceptualized in terms of the discourse set D for any particular discourse, which is the relation space in which we operate at the moment. The agency of the scientist operates non-trivially through R. Such a generalized formulation may be of some help to bring the science and social sciences and humanities discourse nearer with a possibility of mutual understanding. The dropped down dried leaves otherwise would not serve to increase the soil fertility.

Notes

1 For the Tagore-Einstein conversations looked at from various perspectives see the essays in *Materialism and Immaterialism in India and the West: Varying Vistas,* vol. xii, Part 5, ed. Partha Ghose, *Centre for Studies in Civilization in History of Science*, New Delhi: Philosophy and Culture in India, 2010.

A version of the conversations is included in the volume. Besides the conversations appear in book form in *Tagore-Einstein Colloquy*, Rabindra-Bhavan, Visva-Bharati, Santiniketan.

The conversations originally appeared in an appendix of *The Religion of Man* (Macmillan). Details of publication are available in the volume edited by Partha Ghose.

2 See the note at the end of the dialogue in volume ed. Partha Ghose.

3 Letter no. 32, dated February 29, 1912, *cithipatra,* vol. 15, Visva-Bharati Granthan Bivag, Kolkata, sraban, 1402 B S.

The letter is perhaps written to Gourhari Sen, Secretary of Chaitanya Library, See note to letter no. 32 in the above volume, pp. 220. The original letter in reference to which this letter is written appears as letter no. 67, dated shilaidaha, December 9, 1892 in *chinnapatra, rabindra-racanabali,* vol. 11, p. 339, Govt. of West Bengal, Kolkata, August 1989.

4 They could never share their anguish though both had made good friends for a while. The lines are: emoni dui pakhi/ donhare valobase/ tobuo kachenahi pay, which roughly translates to: this is how the two birds love each other, but they do not come nearer.

5 Lectures were made available in book form in *The Religion of Man*. The conversations are now available in *The English Writings of Rabindranath Tagore*, ed. Sisirkumar Das, vol. 3, New Delhi: Sahitya Akademi, 1996. Also reprinted here as Appendix B and C.

6 *The Religion of Man* in *The English Writings of Rabindranath Tagore*, vol. 3, p. 88. The first chapter itself is titled 'Man's Universe'. Apart from the quotations cited, there is a terse formulation like this: 'I do not imply that the final nature of the world depends upon the comprehension of the individual person. Its reality is associated with the universal human mind which comprehends all time and all possibilities of realization. And this is why for the accurate knowledge of things we depend upon Science that represents the rational mind of the universal Man, and not upon the individual who dwells in a limited range of space and time and the immediate needs of life.' (p. 91).

7 Ibid., p. 88.

8 Tony Rothman and George Sudarshan, *Doubt and Certainty*, New Delhi: Scientia, 1998.

9 Francis Fukuyama, *The End of History and the Last Man*, Free Press, 1992. Samuel P. Huntington, *The Clash of Civilizations and the Remaking of the World Order*, Free Press, 1996.
10 C. P. Snow, *The Two Cultures and the Scientific Revolution*, Cambridge University Press, 1959 and *The Two Cultures and a Second Look: An Expanded Version of the Two Cultures and the Scientific Revolution*, Cambridge University Press, 1963.
11 F. R. Leavis, Two Cultures? The Significance of C. P. Snow, *The Spectator*, March 9, 1962. See also Roger Kimball, 'The Two Cultures Today: On the C. P. Snow-F. R. Leavis Controversy', *The New Criterion*, February 12, 1994.
12 Interpretation as is used in German sociological literature is a powerful critique of positivist methodology. It seems to be linked to the Hegelian doctrine of essence. Later instances of this attitude may be found even in Karl Marx and Max Weber.
13 Larry Laudan, *Science and Relativism: Dialogues on the Philosophy of Science*, Chicago: University of Chicago Press, 1990.
14 Werner Heisenberg, Positivism, Metaphysics and Religion in *Physics and Beyond*: *Encounters and Conversations*, trans. Arnold J. Pomerans, Harper and Row, 1971, New York. The Heisenberg statement is: 'The positivists have a simple solution: the world must be divided into that which we can say clearly and the rest which we had better pass over in silence. But can anyone conceive of a more pointless philosophy seeing that what we can say clearly amounts to next to nothing? If we omitted all that is unclear we would practically be left with completely uninteresting and trivial tautologies.'
15 Rabindranath Tagore (Thakur), *Visvaparichay*, 1937, dedication to Satyendranath Bose, dated Santiniketan, 2 aswin 1344 B S, in *rabindraracanavali*, vol. 15, p. 203, Kolkata: Govt. of West Bengal.
16 Einstein's line is: *Raffiniertist der Herr Gott, aberboshaftisternicht.*
17 Rabindranath Tagore, *Creative Unity*, 1922, in *The English Works of Rabindranath Tagore*, vol. 2. See especially the introduction.
18 Stephen Hawking, *The Theory of Everything: The Origin and Fate of the Universe*, Beverly Hills, CA: Phoenix Books, 2006.
19 Ibid., the seventh lecture, pp. 111–24.
20 Ibid.
21 *Kena and Other Upanishads*, Puducherry: Shri Aurobindo Ashram, 2001, p. 5.
22 Stephen Hawking, op cit., p. 124.

References

Das, Sisirkumar (ed). 1994. *The English Writings of Rabindranath Tagore*, New Delhi: Sahitya Akademi, 1996.

Fukuyama, Francis. 1992. *The End of History and the Last Man*, New York: Free Press.

Ghose, Partha (ed). 2010. *Materialism and Immaterialism in India and the West: Varying Vistas*, vol. xii, Part 5, New Delhi: Centre for Studies in Civilization in History of Science, Philosophy and Culture in India.

Hawking, Stephen. 2006. *The Theory of Everything: The Origin and Fate of the Universe*, Beverly Hills, CA: Phoenix.

Heisenberg, Werner. 1971. *Positivism, Metaphysics and Religion in Physics and Beyond: Encounters and Conversations*, trans. Arnold J. Pomerans, New York: Harper and Row.

Huntington, Samuel, P. 1996. *The Clash of Civilizations and the Remaking of World Order*, London: Simon & Schuster.

Kena and Other Upanishads. 2001. Puducherry: Shri Aurobindo Ashram.

Kimball, Roger. 1994. The Two Cultures Today: On the C P Snow-F R Leavis Controversy. The New Criterion, February 12.

Leavis, F. R. 1962. Two Cultures? The Significance of C P Snow. *The Spectator*, March 9.

Rothman, Tony and E. C. G Sudarshan. 1998. *Doubt and Certainty*, Cambridge, MA: Helix Books.

Snow, C. P. 1963. *The Two Cultures and a Second Look: An Expanded Version of The Two Cultures and the Scientific Revolution*, London: Cambridge University Press.

———. 1959. *The Two Cultures and the Scientific Revolution*, London: Cambridge University Press.

Tagore, Rabindranath. 1995. *Chithipatra*, vol. 15, Kolkata: Visva-Bharati Granthan Vibhag.

———. 1937. Visvaparichay. In *Rabindra Racanabali*, vol. 15, 1994, Kolkata: Government of West Bengal.

———. 1922. Creative Unity. In Sisir Kumar Das (ed), *The English Writings of Rabindranath Tagore: A miscellany*, New Delhi: Sahitya Akademi, 1996.

7

WHY THE MOTH CANNOT BE A POET

Saranindranath Tagore

> The mystery of the other is nothing but the mystery of myself.[1]
>
> – Maurice Merleau-Ponty

Preamble

This chapter is concerned with Rabindranath Tagore's understanding of the relationship between reality and human consciousness; further, the chapter formulates a possible answer to the question of how this understanding of the relation contributes to his wide-ranging thinking on the nature of cosmopolitanism. The primary focus of analysis will be the two Tagore-Einstein conversations held during the poet's 1930 visit to Germany that primarily centred on questions of metaphysics and the nature of music. The conversations engage both the just-noted topics of interest, namely the possibilities of human consciousness and the resources of cosmopolitanism. In the first conversation of 14 July (Kaputh) Tagore, in his disagreement with Einstein on the relationship between consciousness and world, developed a conception of the human; and in the second conversation of 19 August (Berlin) a discursive accent was placed on the idea of difference through the specific instance of globally dispersed musical traditions.[2] Both these ideas: the concept of the human and the nature of difference will be important for us to see how cosmopolitanism can be given a foundation allowing for the reconciliation of difference in the possibilities of human consciousness without committing the violence of conceiving unity through exclusion. In order to understand the philosophical contours of the idea of the human in Tagore's conversation with Einstein in particular and his thinking in general I will rely on the phenomenological tradition.[3] I believe that this approach is important to set the philosophical context for Tagore's idea of the human. The human for

Tagore is not so much an invocation to refute metaphysical realism, but is a sign under which the richness of experience can be unified. Thus the human as the limit-range of experience under whose sign unity can be addressed emerges with a fundamental ethical resonance. The task of this chapter is to delineate how this idea of the human, which so forcefully emerges in the poet's conversation with Einstein can fundamentally motivate the ethical concern that I will be calling cosmopolitan.

On difference and music, 19 August 1930, Berlin

The conversation between Tagore the composer and Einstein the violinist concerning the nature of music records one of the most lucid characterizations of musical difference globally conceived. The topic of music is introduced by Tagore in response to Einstein's reworking of the realist agenda carried over from the earlier conversation focusing on themes metaphysical. Einstein observes that even in the world of elementary particles statistical order prevails, profiling the well-known classical insistence that quantum mechanics at least in its Copenhagen incarnation is incomplete. Tagore responds by stating that it is the constant harmony of chance and determination which makes the 'drama of existence' escape the 'desultory' landscape of Being. Musical experience enters the discussion at this point when Tagore summons the practitioners of the *rāga* tradition to illustrate the weaving of 'chance' and 'determination':

> Our composers give a certain definite outline, a system of melody and rhythmic arrangement and within a certain limit the player can improvise upon it. He must be one with the laws of that particular melody and then he can give spontaneous expression to his musical feeling within the prescribed regulation. We praise the composer for his genius in creating a foundation along with a superstructure of melodies, but we expect from the player his own skill in the creation of variations of melodic flourish and ornamentation.[4]

The duality between freedom and necessity which Einstein attempts to harmonize through the idea of statistical order, Tagore likens to the melodic freedom in the performance-structure of a classical Indian musician working within the strict parameters of a given *rāga*-rule. The idea of the *rāga* provides the thread linking the metaphysical underpinnings of music and physics. Importantly for us, though, the duality between necessity and freedom profiles a significant horizon of difference between the European and the Indian traditions. The freedom of Sviatoslav Richter playing a Bach variation is fundamentally different from the freedom articulated in a Nikhil Banerjee invocation of the raga *Hemant*. This is an important difference that should interest the comparative musicologist but need not detain us here for long. Implicitly, invoking the idea of a *tāla*, Tagore fluently enunciates this difference between the two accounts of musical performance: 'In European music you have a comparative liberty about time but not about melody, but in India we have freedom of melody but no freedom of time.'[5] The two traditions, at least in the aspect of performativity, in this formulation, stand in inverse relationship to each other with respect to the two fundamental aspects of music: melody and time. Of course, many other points of difference, essentially inscribed can be invoked: the singular melodic line microtonally developed in the Indian tradition stands apart from the contrapuntal development of melodic multiples in the European tradition can be added to the continuous curvature of the musical flow in the Indian tradition versus the linearity of European music, chamber or symphonic. Without proceeding with this comparative account of music, the philosophically motivated example is established: music-history provides an example of difference of seeming incommensurability, which makes it difficult to observe the Tagorean theme: anything that is human instantaneously becomes ours in enjoyment, wherever its origin might be.[6] Further, in the context of difference bordering on the incommensurable, of which the instance of music provides a particularly salient example, a distinction between the human and the monstrous emerges, a theme that will play an important role in the remarks to follow. For now we can note that I define *the monstrous to be the field that is incommensurable to the human*. Returning to the question of music, the issue of incommensurability is well captured in the musical discussion because unlike other cultural formations, difference here is so deeply anchored that a musical expression in one tradition may not even appear as music when heard by a listener nurtured in another tradition. I make this suggestion tentatively, because the microtonal development and contrapuntal construction are not available to both the musical traditions together; and further I have some anecdotal evidence in support of this claim. At any rate, the main point is that musical difference motivates the issue of cosmopolitanism in Tagore's worldview because one of the great problems of his life-work centred on the question of how to think of unity within a world of great difference. This problem implicitly surfaces in this wonderful discussion of musical difference.

How to think about unity where the trope of incommensurability threatens rupture? This is a problem for cosmopolitanism, at least in the cultural sphere, because without the traces of commensurability the determining force of tradition cannot be overcome in the welcoming of difference. In these remarks I wish to profile the idea of the human as playing a central role in regard to this strategy of disclosing commensurability; and needless to say that the focus on the human boldly emerges in Tagore's argument against Einsteinian realism to which I now turn. Musical difference is *difference within the human* and cannot summon *the incommensurable* implicit in all discourses of the fragment. A proper understanding of the incommensurable should invoke the beyond of the human – the monstrous. Such a difference perhaps is evident in an implicit fashion in Tagore's invocation of the human in the Kaputh conversation on the nature of reality.

On the human, 14 July 1930, Kaputh

The Kaputh conversation, occurring before the exchange on music, ostensibly about the nature of reality as it surfaces in the philosophy of physics registers a fundamental disagreement about metaphysical orientation between the poet and the physicist. In an act of remembrance, Tagore recounted later the substance of their philosophical parting:

> I could see that Einstein held fast to the extra-human aspect of Truth. But it is evident to me that in human reason facts assume a unity of Truth which is only possible to a human mind. While logic belongs to man's mind, I can conceive of another field of logic for another type of mind other than ours. So, when Einstein talked to me of Pythagorean geometry, I could conceive of an existence where space is not the factor but where Time reigns as it does in music. It is not impossible that a mind can live in time and have no notion of space. For such a mind, existence might be the kind comprehended in music.[7]

Tagore's point here may be best explicated through the lens of Kantian philosophy, which momentously introduced in philosophical discourse the idea that human consciousness is a constituting agent of experience. Kant, initiating what is sometimes called a Copernican revolution in the philosophy of cognition, argued that the world-experience (*phenomena*) is not merely a datum of receipt from an already-formed world, which corresponds to a cognitive interiority, but is rather a constituted product of consciousness. In other words, the very process of cognition alters the world as it is in itself (*noumena*) to produce the object of experience through the imposition of forms and categories belonging to the experiencing consciousness. Thus emerges the unbridgeable divide between the world as it appears to human consciousness and the world as it is in itself (*noumena*) in Kant's thinking. Note that the Kantian position does not retreat into pure idealism because consciousness *qua* human consciousness brings under its sweep the *whole* possibility of a world. And this range of possibility exhausts the intentional reach of the human.[8] The beyond of the human is monstrous in the vocabulary adopted in this chapter. Thus in the transcendental thrust of the Kantian approach, the constituting aspect of human consciousness, like the eye of the visual field to use a Wittgenteinian metaphor, must necessarily be transcendentally related to the world-experience it constitutes.[9] Space and Time, to continue with the gloss, as modes of intuition make up the form of the world-experience, whereas a series of categories such as causality structures the relationship between the entities comprising that world. The main point here for us is that in the Kantian approach a distinctive conception of the human emerges that constructs a vast field of intersubjectivity where all possible objects of cognition (intentional objects for later phenomenologists) are rendered commensurable through the idea of consciousness *qua* human consciousness. In this reading, the incommensurable, like Tagore's wonderfully wrought example of purely temporal experience, *not apparent but absolute*, is relegated to the domain of the monstrous.

It is possible to read Tagore's disagreement with Einstein in such broadly Kantian-phenomenological terms because it shows the great philosophical coherence of the fundamental position that he is advancing in the conversation. The conception of the human that he is defending I term phenomenological because a good case can be made that Kant is the progenitor of this school of philosophy because Husserl will proceed later to develop the main tenets of phenomenology by leveraging on this Kantian move to the subject by arguing that philosophical (phenomenological) reflection should focus on the immediate resources of consciousness and set aside in an act of *epoché* the world that is transcendent to the self. I am proposing that the fundamentally Kantian idea of the human resurfaces in Husserl because a monstrous phenomenology would be inaccessible to the human. Again, in this reading the human is summoned within the range of phenomenological resonance. A phenomenologically informed reading of the human would think the problem of the commensurable within the limits of the human and would relegate the incommensurable to the monstrous. This thought needs further probing.

One great difference between Kant and Husserl that is of particular salience for these remarks is a point raised by J.N. Mohanty to the effect that Husserlian phenomenology can better accommodate difference than Kant's philosophy.[10] This is so because for Kant, supposing a Newtonian conception of the physical universe, can only posit a classical framework wherein space is wholly Euclidean and time is Newtonian. Thus it can be argued that the Kantian account of world-experience may not travel too well to a scientific domain where non-Euclidean space-formations scaffold relativistic cosmology. Husserl does not fall prey to a problem of this nature, Mohanty argues, because phenomenological reflection sees constitution in terms of meaning (*sinn*) formations rather than Kant's pure concepts of the understanding. If the function of constitution is meaning-formations then the field of plurality opens up because the summation of all intentional objects emerge in consciousness as a series of meanings thus sidestepping the problem of legitimation where one theoretical posture trumps another. Indeed, Husserl can attend to meaning of intentional objects because through the bracketing of transcendent (worldly) entities he *qua* meditating consciousness can focus on the world as it emerges for the phenomenological subject. The question of the external world is bypassed because it is not subjected to Cartesian scepticism but rather ontological commitment towards the external world is neutralized. It is only after this process of bracketing (*epoché*) that the transcendent world can appear as meaning structures in the interiority of consciousness purified of externality. All postures, theoretical or otherwise, have equal respect because the fundamental proposal of Husserlian phenomenology is that the task of consciousness is meaning constitution. In such a scenario, the monstrous would become incommensurable to the human only within the intuition, in consistency with both Kant and Husserl, which keeps open the possibility of experiences that are beyond the limits of human possibility. A monstrous phenomenology, while possible, riding on the back of absolute difference, would be incommensurable to the human. This phenomenological framework, deriving from Kant and Husserl, can service Tagore's position against Einstein very well:

> It is not difficult to imagine a mind to which the sequence of things happens not in space but only in time like the sequence of notes in music. For such a mind its conception of reality is akin to the musical reality in which Pythagorean geometry

> can have no meaning. There is the reality of paper, infinitely different from the reality of literature. For the kind of mind possessed by the moth, which eats that paper, literature is absolutely non-existent, yet for Man's mind literature has a greater value of Truth than the paper itself.[11]

At this juncture the phenomenological formulation of the human sounds its arrival:

> If there be some Truth which has no sensuous or rational relation to the human mind it will ever remain as nothing so long as we remain human beings.[12]

The phenomenologically motivated reading suggests that the philosophical shape of Tagore's formulation escapes Einstein who is endorsing the classical view of physical structure when he responds thus: 'then I am more religious than you are.' Einstein invokes religion in this context because he thinks that theism must support a reality that is absolutely independent of the intuitions of consciousness. Tagore is implicitly appealing to the monster in the suggestion that a 'truth' incommensurable to the total possible reach of human consciousness is not a truth in so far as we are human. Alternatively, in phenomenological terms, an object that in principle cannot emerge in the structure of intentionality cannot *possibly* be an object for that structure. Though Tagore makes the remark about science it is equally true of religion: 'this world is a human world – the scientific view of it is also that of the scientific man.'[13] We cannot disown our consciousness and assume the structure of any monstrous consciousness. Such a resistance to Einsteinian realism does not argue against the possibility of a world free of human consciousness but makes it irrelevant for a consciousness that is human. Following the phenomenological-Kantian hermeneutic on offer, a profitable interpretation of Tagore's conception of the 'eternal man' can be developed here. Tagore remarks:

> The world apart from us does not exist; it is a relative world, depending for its reality upon our consciousness. There is some standard of reason and enjoyment which gives it Truth, the standard of the Eternal Man whose experiences are through our experiences.[14]

The idea of the 'Eternal Man' is invoked to blunt the charge of anti-realism (at least Einstein's conception of it) suggested by the idea of the relative world. If the notion of 'Eternal Man' is read phenomenologically it would be taken as signifying the possible span of human experience that the individual can range over. In other words, the world is not relative to individual humans but to the entire range of possible human experiences. Tagore, in my estimation makes exactly this point, when he remarks:

> humanity is composed of individuals, yet they have their interconnections of human relationship, which gives living solidarity to man's world. The entire Universe is linked up with us, as individuals, in a similar manner, it is a human universe.[15]

A broadly similar account of Tagore's characterization of the relationship between world and consciousness can be issued from the perspective of the *Vedāntic* idea of *māyā* (mentioned later) but, in the current context, the phenomenological hermeneutic is preferred given the fact that in this instance Tagore's interlocutor is Einstein who is well-familiar with the Germanic intellectual tradition. Thus in my view, Einstein's disagreement with Tagore does not merely reflect, as is usually supposed, his resistance to the Copenhagen framework of quantum physics but mirrors a broader disagreement with a central line of European philosophical reasoning beginning with Kant and continuing with Husserl that refuses to purchase realism at least of the Einsteinian sort.[16] Alternatively, if not a disagreement, at least one can suggest that Einstein's responses to Tagore's philosophical formulations do not seem to engage this central Germanic tradition. This observation should emerge as somewhat of a mystery to a reader of this conversation.

Beyond the conversation: an excursus on unity

In order to see how such a phenomenologically construed account of the human can foundationally motivate a cosmopolitan gesture, the relationship between the human and the incommensurable, somewhat latent in the mention of the monstrous, may be further discussed. The idea of unity is important in this regard because if a limiting line can be drawn around the human as a possible range of experience, then the concept (human) posits a unity. Consequently, as an implication of the phenomenological analysis, the shapes of difference that inhabit the interiority of this limiting field must in principle be commensurable. Further, needless to say, the possibility of the cosmopolitan attitude that seeks to include the multiple in the formation of subjectivity, must posit commensurability. In my view, it is in the sense of this argument, that the idea of the human as a limit-concept provides grounding for Tagore's philosophy of cosmopolitanism. The conversation with Einstein provides a platform on which this idea is given a philosophical airing.

The argument needs elaboration in the particular context of Tagore's philosophical texts in order to show that his conversation with Einstein systematically extended his cosmopolitan concerns. Let us move in this direction by attending to the remarkable metaphor of the moth. Tagore makes the point that for the moth that eats the paper literature is non-existent; only paper is real for a moth-like mind. In this regard, where literature is concerned, the moth-mind assumes the role of the monster standing in a relation of incommensurability with the human mind.[17] No human mind can assume a monstrous shape in relation to another human mind because the phenomenologically driven point is that such a consciousness features the enabling conditions that draw the limits of the possible range of meaning-formations. Meanings that are monstrous are *outside* the orbit of this limiting line. The apparent incommensurability between two kinds of musical experience is resolved in this view because they can assume musical meaning *within* the limit of the human. Whether it actually does in a particular consciousness is not the issue but that it possibly can in *some* consciousness shows that neither of these traditions can be consigned to the realm of the monstrous from the perspective of the other. In other words, it is the human as a range of possibilities *qua* human possibilities that directs diverse experiences, seemingly incommensurable, towards the horizon of unity or at least the domain of the commensurable. In this view, then, the human, phenomenologically understood, forwards an argument for commensurability. All differences, musical or otherwise, can be mediated without subverting the registry of difference, in the range of possibilities that a type of consciousness, in this instance, human, profiles. Beyond the human, the region of absolute incommensurability opens.

The cosmopolitan agenda so much a part of Tagore's achievements cutting across several fields of expression surfaces in the conversations under discussion. Indeed, it is the phenomenological account of the human which is brought to bear upon the possibility of unity:

> The species . . . belongs to a unity, to humanity. Therefore the entire human mind realizes Truth; the Indian or the European mind meet in a common realization.[18]

The idea is that once the human *qua* consciousness is seen as the signpost of commensurability, all differences, Indian and European in this instance, move towards a horizon of unity. In Tagore's phenomenology of the human, this formation of unity made possible through the commensurable, is precisely what is impossible when the monstrous or the moth-mind is invoked. Such an account can deeply motivate a cosmopolitan attitude and a cosmopolitan community (ethos). The notion of a cosmopolitan community understood in terms of an ethos further pushes the notion of the human towards the direction of a normatively regulated social order where it merges in an intersubjective grid summoning the other without whom the ethical mode would remain unmotivated. In this regard the notion of ethos is best presented through Michel Foucault's formulation, which masterfully captures the intertwining relationship between attitude, modernity and ethos:

> I wonder whether we may not envisage modernity rather as an attitude than as a period of history. And by 'attitude', I mean a mode of relating to contemporary reality; a voluntary choice made by certain people; in the end, a way of thinking and feeling; a way too of acting and behaving that at one and the same time marks a relation of belonging and presents itself as task. A bit, no doubt, like what the Greeks called an ethos.[19]

In order to see the connecting tissue between the human and the cosmopolitan some thought must be given to how Tagore envisioned cosmopolitanism.

Tagore's understanding of the human embraced a cosmopolitan account of unity. In his vision, as I have been arguing, the commensurable is a feature of the human landscape, otherness as the signature of the incommensurable is monstrous defined as lying outside the possible range of human experience. Thus the unity of the human is to be profiled over and against the alterity of the monstrous. This is how, in my view, Tagore's point made to Einstein noted earlier that 'the species . . . belong to a unity, to humanity' is to be understood. However, the issue of difference within the human cannot be elided because the very idea of cosmopolitanism arises within the question of how to negotiate difference. In other words, difference within the cultural sphere cannot be altogether erased through the invocation of the monstrous as a limit-concept. Thus in the context of a phenomenologically driven account of the human, *difference within the limit of the human* must be distinguished *from the absolute incommensurability of the monstrous*.

Tagore's account of unity, which distinguished between two senses of the term, is instructive in the current context.[20] Two remarks from Tagore on unity will have to suffice. First:

> When the science of meteorology knows the earth's atmosphere as continuously one, affecting the different parts of the world differently, but in a harmony of adjustments, it knows and attains truth. And so too, we must know that the great mind of man is one, working through the many differences which are needed to ensure the full result of its fundamental unity. When we understand this truth in a disinterested spirit, it teaches us to respect all the differences in man that are real, yet remain conscious of our oneness; and to know that perfection of unity is not in uniformity, but in harmony.[21]

Second:

> Uniformity is not unity. Those who destroy the independence of other races, destroy the unity of all races of humanity. Modern Imperialism is that idea of Unity, which the python has in swallowing other live creatures . . . when we respect the true individuality of men, then we can discover their true unity.[22]

Tagore is distinguishing between unity as uniformity and unity as harmony and he rejects the former and accepts the latter on normative grounds. Unity as uniformity is achieved through exclusion whereas unity as harmony pivots on an open conception of human identity that always seeks encounter with alterity. Tagore's conception of unity as harmony implies an education of attitudes (*bhābsikkha*)[23] and implies a path towards cosmopolitanism that Derrida, following Levinas, would follow later:

> Interruption of the self by the self, if such a thing is possible, can or must be taken up by thought: this is ethical discourse – and it is also, as the limit of thematization, hospitality.[24]

If cosmopolitanism is derived from the universality of reason as Martha Nussbaum tends to do, then there is a danger, as some of her critics point out, of flattening out difference. However, if difference is to be preserved within a Tagorean schema that sees the human under the flag of unity, then hospitality would appear as a better source for the idea of cosmopolitanism. Hospitality then is interruption of self by self, wherein identity is never frozen into a final shape but is essentially fluid. Auto-interruption is the condition of possibility for hospitality. There is one remark by Tagore in a speech delivered in Tokyo where he explicitly connects hospitality to the cosmopolitan gesture:

> We should have faith in ideals that have matured in the spiritual field through ages of human endeavor after perfection, the golden crops that have developed in different forms and in different soils but whose food value for man's spirit has the same composition. These are not for the local markets but for universal hospitality, for sharing life's treasure with each other and realizing that human civilization is a spiritual feast the invitation to which is open to all, it is never for the ravenous orgies of carnage where the food and the feeders are being torn to pieces.[25]

Indeed, this sense of the hospitable animated much of Tagore's social and political thought and in my view has far-reaching philosophical resonance. For instance, it can be seen as blunting the postmodern assertion of Lyotard that all unity-constructions, *grand narratives*, are essentially self-legitimating and thus hegemonic through exclusion:

> I will use the term *modern* to designate any science that legitimates itself with reference to a meta discourse . . . making an explicit appeal to some grand narrative, such as the dialectics of Spirit, the hermeneutics of meaning, the emancipation of the rational or the working subject, or the creation of wealth.[26]

In this reading of modernity, humanism that is underwritten by the assertion of unity would emerge as hegemonic because it would be constituted through exclusion. However, the point that needs to be made is that the narrative of unity that draws its inspiration from the auto-interruption of self, the condition of possibility for hospitality, cannot be subjected to the same criticism. How this alteration would affect the philosophical investigation of modernity remains an open question that will not detain us in the present exercise. It will suffice to note that Tagore himself thought that the modernity of his great predecessor Rammohan Ray was built on this other cosmopolitan conception of the modern:

> His attempt was to establish our peoples on the full consciousness of their own cultural personality, to make them comprehend

> the reality of all that was unique and indestructible in their civilization, and simultaneously, to make them approach other civilizations in the spirit of *sympathetic co-operation*.[27]

Tagore's conception of sympathy, again possible only within a field of commensurability, ultimately motivates a cosmopolitan gesture moved by hospitable consciousness. The relationship between world as a system of difference and subjectivity is one of welcome, wherein the possibility of unity is profiled, through the attitude (*bhāb*) of hospitality. However, the idea of the interrupted self does not travel in a free-floating fashion but is rooted in the contours of subjectivity. It is only that the world as a system of difference, if it is to be welcomed, must be ushered into a home, or in phenomenological terms, into subjectivity. Tagore's point is that this subjectivity, phenomenologically construed, is open to the entire range of the human and is not frozen in a reified shell. Thus, the root-subject, while it has a foundation in an inherited tradition, for the cosmopolitan narrative, journeys through possibilities, shaped and reshaped, and ultimately flourishes within the range of the human, which is a field of unity containing differences.

Tagore commented in the *Religion of Man*:

> I do not imply that the final nature of the world depends upon the comprehension of the individual person. Its reality is associated with the universal human mind which comprehends all time and all possibilities of realization.[28]

The universal human mind that Tagore is referring to here and also in the conversation with Einstein is the comprehending subject of 'all possibilities of realization.' In other words, the individual mind is a limit which travels within the range of the human that in its totality is an object of comprehension of the universal mind. In this reading, the human that is both limiting and transcending, poetically celebrated by Tagore through many names (*aparājita mānush*, unvanquished human, and *mahā mānab*, over-human), is at once the total range of possible experience *and* its limit-condition.[29] This idea is stressed, I think, in the conversation with Einstein when Tagore remarks: 'Truth is the perfect comprehension of the Universal Mind. We individuals approach it through our accumulated experience . . . how otherwise can we know Truth?'[30] From a Kantian-phenomenological angle, Tagore's universal mind can be understood in terms of the idea found in both Kant and Husserl (albeit in different ways) that actual individual experience emerges as a limit to the entire possible range of experience that consciousness *qua*human consciousness is capable of. Of course, such an extension has to be drawn rather carefully because what is being stressed here is the human as a limit-range and not the *constituting function* of subjectivity which is so central to both Kant and Husserl. The constituting function of transcendental subjectivity cannot be exactly read into Tagore's conception of the human but that is not a loss for this phenomenological reading. The point that is immediately important is that the idea of hospitality extends this idea of difference within the range of the human to areas of normative concerns, which I am calling cosmopolitan.[31] I have just considered, albeit all too briefly, some of these considerations.

Back to the Kaputh conversation: on cosmopolitanism and the human

Tagore shows how this fundamentally metaphysical conception of subjectivity can service the cause of cosmopolitanism. The discussion so far, leveraging on the phenomenological tradition, was designed to substantiate in as much detail as possible a fundamental distinction evident in Tagore between difference within the human and the incommensurability of the monstrous. The idea of unity both in its abstract metaphysical incarnation and in its ethico-practical comportment enshrined in the gesture of welcome is important for the philosophical construction of cosmopolitanism. One of the sources of cosmopolitan ethics that seeks unity across cultural traditions has been sourced from the Enlightenment conception of human dignity and universal reason.[32] However, it can be argued that if cosmopolitanism is to value the particularities of human expression wherever they might have their origins, then it cannot overlook the impress of inherited tradition by anchoring itself in the transcendental horizon of abstracted reason. This particular issue concerning the grounding of cosmopolitan ethics is perfectly mirrored in the postmodern construction of modernity alluded to earlier. Lyotard argues that modernity, as a complex of legitimating narratives, is violent precisely because it subverts difference through the construction of unity through eschewal. This point sheds light on the issue of the source of cosmopolitanism. If meta-narratives are violent towards difference, then the idea of human dignity, which is untouched by the particularities of tradition, cannot serve a cosmopolitan ethos. However, the postmodern solution to the problem of modernity is found in the location of little narratives, which, like the structure of a game, are consolidated by their own rules. To *be* is to belong to such a narrative incommensurably related to other narratives. In such a view, difference is preserved but at the expense of hospitality. In postmodernism, perhaps the respect for difference at best can emerge in the value of toleration but cannot sustain the interruptible self as the source of hospitality. The philosophical point is, suspending the labels for a moment: neither the universalist nor the particularistic standpoints can motivate the cosmopolitan gesture. The former effaces whereas the latter seals the shapes of particularities. For the former there is no home, for the latter there is no world.

In the idea of the human, the home and the world are preserved. Tagore's great evocation of the human, the cornerstone of his conversation with Einstein, serves as the regulating principle of his philosophical interventions ranging over many issues. A fundamental question for him concerned the problem of unity. Without the possibility of unity, he realized, anticipating the postmodern *aporia*, one cannot reach beyond the confines of inherited tradition. However, this concept of unity if it is to serve a cosmopolitan cause cannot operate as a legitimating function. Thus he distinguished between the two senses of unity: unity as uniformity and unity as harmony, again anticipating the postmodern linkage of unity with violence. At this juncture, the idea of the human as envisaged by Tagore registers itself by announcing the distinction between the difference within the human and the absolute incommensurability of the monstrous. Narratives, in so far as they are human, are open to other narratives under the sign of unity. The cosmopolitan attitude (*bhāb*) then is always manifested in the line towards unity within the range of the human. Tagore's conception of the human engages both difference and unity in an exceptionally creative way to answer how hospitable consciousness can sustain cosmopolitanism.

The relationship between the human and the cosmopolitan was a part of Tagore's conversation with Einstein as is evident in the remark brought up earlier:

> The species . . . belongs to a unity, to humanity. Therefore the entire human mind realizes Truth; the Indian or the European mind meet in a common realization.[33]

It is interesting to note that this formulation of cosmopolitanism invoking the idea of the unity of consciousness as the total range of possible experience (phenomenologically interpreted) also emerges in the Kaputh conversation out of an appeal to *Vedānta*. Here is the full context:

> According to Indian Philosophy there is Brahman, the absolute Truth, which cannot be conceived by the isolation of the individual mind or described by words, but can only be realized by completely merging the individual in its infinity. But such a Truth cannot belong to Science. The nature of Truth which we are discussing is an appearance – that is to say what appears to be true to the human mind and therefore is human, and may be called Maya or illusion.[34]

As Tagore in this instance is appealing to the ineffability of Brahman and is distinguishing between the *paramārthika* (absolute truth) and *vyāvahārika* (mundane truth) where cognitions of appearance inclusive of scientific claims have their place, it is safe to assume that he has Sankara (*Advaita*) Vedānta in mind. This *Vedāntic* view also invests in the idea that the field of intentional experience is to be understood as the total possible range of human experience and Tagore clearly makes this point when he remarks that the species (the human) belongs to a unity, echoing Einstein's gloss of the *Vedāntic* view that *māyā* is not the illusion of the individual but of humanity as a whole.[35] Parenthetically, indeed a case can be made that in the articulation of the sublatable field of *māyā* Sankara vitally anticipates the phenomenological characterization of consciousness as intentional.[36] It is only that in Sankara, unlike Husserl, intentionality is neutralized in the final order of monistic experience. In so far as Tagore relates his view of unity to *Vedānta*, it can be claimed as an extension that cosmopolitanism as he envisioned it as a task of the human can be derived from the history of Indian philosophical concerns as well. Such a position indicates the point of interest, within the context of these remarks, that a dialogue can be established between *Vedānta* and phenomenology on wider issues of ethics and social philosophy but this is a line of comparative reasoning that must await another occasion. For now, the only issue that needs to be stressed is that the *Vedāntic* and the phenomenological hermeneutic of the human in Tagore are indeed not as foreign to each other as one might suppose.

Unlike the apathy inscribed in the postmodern attitude and the subtractions of the modern, for Tagore, expressions as products of consciousness *qua* human consciousness can never be situated within the fragment. His conception of the human shows, pivoting always on hospitable consciousness, exactly how an account of unity that is not violent can be philosophically mounted. We started with music as an example of difference within the human, but expressions can be multiplied to cover the entire range of human possibilities that are at once possibilities of experience working toward what Tagore would call 'meeting of realizations.' If narratives of unity are modern in their legitimating drive as the postmodern would have it, one can perhaps argue that Tagore is moving towards a conception of the modern, which in its cosmopolitan reach inaugurates a new mode of thinking on the problem of modernity. And if this is right, perhaps an illegitimate slide is evident in the postmodern incredulity towards narratives of unity, which is arrived at through the inspection of a *specific* account of the modern. In these remarks, I have considered a few elements constitutive of a larger task of showing how this alternative account of modernity, derived from the Indian philosophical history but in conversation with phenomenology, can be grounded in a conception of the human, which is the key-note of Tagore's conversation with Einstein.

Tagore's humanism teaches that the ethics of toleration can be extended to the incommensurable field of the monstrous but can never be sufficient in framing the meeting of the human self with the human other.[37] Tagore's understanding of the meaning of the human seeks its sustenance in the gesture of welcome, beyond toleration, wherein the contours of consciousness is kept infinitely open to the mystery of the other.

Notes

1 Maurice Merleau-Ponty, *The Prose of the World*, Evanston, IL: Northwestern University Press, 1973, p. 135.

2 I will refer to these conversations throughout this essay as they appear in the combined *The Kenyon Review* and *Stand* magazine's volume celebrating the centennial anniversary of the Nobel Prizes published in conjunction with the Nobel Museum. *The Kenyon Review*, Volume XXIII, no. 2, Spring, 2001 and *Stand*, Volume 2(4)/3(1), March 2001, pp. 20–30. Hereafter this publication will be referred to as *Kenyonstand*.

3 There is some discussion of Tagore's philosophical orientation (in the Tagore-Einstein conversation) in relation to Hilary Putnam's internal realism, but in the current view the phenomenological treatment hold greater promise in showing how this metaphysical conception of the human and its epistemic consequences can be extended to broader social concerns. I will be including Kant in the general discussion of phenomenology because, following Husserl, I take Kant (along with Hume) as one of the major harbingers of phenomenology. There is of course no evidence to suggest that Tagore himself was influenced directly or indirectly by this particular strand of Western philosophy. For a discussion of the Putnam position in relation to Tagore's conversation with Einstein see Dipankar Home and

Andrew Robinson, 'Einstein and Tagore: Man, Nature and Mysticism', *Journal of Consciousness Studies*, 2(2) (Summer 1995): 167–79. Amartya Sen also refers to Putnam's views on realism in 'Tagore and his India', in *The Argumentative Indian*, London: Allen Lane, 2005, p. 104.

4 *Kenyonstand*, p. 27.

5 Ibid., p. 28.

6 'Whatever we understand and enjoy in human products instantly become ours, wherever they might have their origin. I am proud of my humanity, when I can acknowledge the poets and artists of other countries as my own.' *Letters to a Friend*, New Delhi: Rupa, 2002, p. 111.

7 Rabindranath Tagore, 'My Memories of Einstein', in *Kenyonstand*, p. 19.

8 Throughout these remarks I will be using the term intentionality (intentional) in its philosophical usage, particularly as used by the phenomenological tradition. In this restricted sense, to say that consciousness is intentional is to simply mean that consciousness has the property of relating to an object. For Husserl, the intentional object may be a worldly or a fictitious object, it simply has to emerge as an object for consciousness. Phenomenologically oriented philosophers usually take intentionality to be an *essential* feature of consciousness.

9 See Ludwig Wittgenstein, *Tractatus Logico-Philosophicus*, London: Routledge & Kegan Paul, 1961, p. 57.

10 See J. N. Mohanty, *The Possibility of Transcendental Philosophy*, Dordrect: Martinus Nijhoff, 1985. See especially the discussion in pp. xxv–xxvi, where the limit of relativism is phenomenologically negotiated to show how the pluralist point of view can be advanced.

11 *Kenyonstand*, p. 25.

12 Ibid.

13 Ibid, p. 22.

14 Ibid.

15 Ibid.

16 While in Kant one has a defence of transcendental idealism, for Husserl ontological commitment to a mind-transcendent world is neutralized in the act of bracketing which allows phenomenological reflection to focus on the evidence of consciousness purified of mundanity.

17 Here is what Tagore says to Einstein on the moth-mind: "There is the reality of paper, infinitely different from the reality of literature. For the kind of mind possessed by the moth, which eats the paper, literature is absolutely non-existent, yet for Man's mind literature has a greater value of Truth than the paper itself. In a similar mind. If there be some truth which has no sensuous or rational relation to the human mind it will ever remain as nothing so long as we remain human beings." *Kenyonstand*, p. 25.

18 *Kenyonstand*, p. 23.

19 Michel Foucault, 'What Is Enlightenment', in Paul Rabinow (ed), *The Foucault Reader*, New York: Pantheon Books, 1984, p. 39.

20 For my earlier account of Tagore's conception of cosmopolitanism shorn of the current worry about the ways in which the human operates as a grounding concept see Saranindranath Tagore, 'Tagore's Conception of Cosmopolitanism: A Reconstruction', *University of Toronto Quarterly*, 77 (Fall 2008), pp. 1070–83.

21 Rabindranath Tagore, *Creative Unity*, New Delhi: Rupa, 2002, pp. 171–2.
22 Rabindranath Tagore, 'The Union of Cultures', in Sisir Kumar Das (ed), *The English Writings of Rabindranath Tagore*, New Delhi: Sahitya Akademi, 1999, p. 435.
23 Here I have the following remark from Tagore in mind. In *Shikkar Herpher* he writes (offered here in my translation): 'If from childhood one is educated along with language instruction about attitude formation (*Bhābshikkhā*) and if one's entire life is balanced with this attitude, then only harmony can be established in life; we can then live naturally and can apprehend the true measure of things.' Shikhār Herpher, Kolkata: Visva Bharati Press, 1992, pp. 13–14.
24 J. Derrida, 'A Word of Welcome', *Adieu to Emmanuel Levinas,* Stanford: Stanford University Press, 1999, p. 51.
25 Rabindranath Tagore, 'The Ideals of Education', in *The English Writings of Rabindranath Tagore*, New Delhi: Sahitya Akademi, p. 613.
26 Jean-François Lyotard, *The Postmodern Condition*, Minneapolis: University of Minnesota Press, 1984, p. xxiii.
27 Rabindranath Tagore, 'Rammohun Roy', in *The English Writings of Rabindranath Tagore*, New Delhi: Sahitya Akademi, vol. 3, p. 667. Emphasis added.
28 Rabindranath Tagore, *Religion of Man, in The English Writings of Rabindranath Tagore*, New Delhi: Sahitya Akademi, vol. 3, p. 91.
29 The first formulation appears in *Crisis in Civilization* and the second in *Shesh Lekha.*
30 *Kenyonstand*, p. 23.
31 Can one in this context invoke Heidegger and suggest that the Tagore's conception of the human would travel better in the Heideggerian direction of *Dasein* which is not transcendentally formulated? The idea that phenomenological reflection is specific to a structure of consciousness that is inherited by the subject of phenomenological reflection is a Cartesian supposition that must be shared by all phenomenologists, transcendental or existential. For instance, when Merleau-Ponty extends his phenomenological analysis to corporeality, the human body as a limit-range must be operant because Merleau-Ponty *qua* phenomenologist has no access to monster-bodies.
32 See for instance Martha Nussbaum, 'Patriotism and Cosmopolitanism', in Joshua Cohen (ed), *For Love of Country*, Boston: Beacon Press, 1996, pp. 3–17.
33 *Kenyonstand*, p. 23.
34 Ibid.
35 See Ibid, p. 23.
36 For a detailed phenomenological treatment of the *Vedāntic* concept witness consciousness (*sākṣin*) see Bina Gupta, *The Disinterested Witness: A Fragment of Advaita Vedānta Phenomenology*, Evanston, IL: Northwestern University Press, 1998.
37 Such a conception of the human as developed in these remarks can be rendered consistent with the idea that humans have duties going beyond toleration towards non-human animals (well-being) and the environment (sustainability), both close to Tagore's ethical concerns.

References

Das, Sisir Kumar (ed). 1999. *The English Writings of Rabindranath Tagore*, New Delhi: Sahitya Akademi.

Derrida, Jacques. 1999. *Adieu to Emmanuel Levinas*, Stanford: Stanford University Press.

Foucault, Michel. 1984. What is Enlightenment. In Paul Rabinow (ed), *The Foucault Reader*, New York: Pantheon Books.

Gupta, Bina. 1998. *The Disinterested Witness: A Fragment of Advaita Vedānta Phenomenology*, Evanston: Northwestern University Press.

Home, Dipankar and Andrew Robinson. 1998. Einstein and Tagore; Man, Nature and Mysticism. *Journal of Consciousness Studies*, 2(2)(Summer 1995).

Lyotard Jean-François. 1984. *The Postmodern Condition*, Minneapolis: University of Minnesota Press.

Mohanty, J.N. 1985. *The Possibility of Transcendental Philosophy*, Dordrecht: Martinus Nijhoff.

Nussbaum, Martha. 1996. Patriotism and Cosmopolitanism. In Joshua Cohen (ed), *For Love of Country*, Boston: Beacon Press.

Sen, Amartya. 2005. *The Argumentative Indian*, London: Allen Lane.

Tagore, Rabindranath. 2002. *Creative Unity*, Kolkata: Rupa.

———. 2002. *Letters to a Friend*, Kolkata: Rupa.

———. 1992. *Shikhār Herpher*, Kolkata: Visva Bharati Press.

Tagore, Rabindranath and Albert Einstein. 2001. The Tagore-Einstein Conversation. *The Kenyon Review*, XXIII(2), and Stand, 2(4)/3(1). (Published in a combined volume to honour the centennial celebration of the Nobel Prizes.)

Tagore, Saranindranath. 2008. Tagore's Conception of Cosmopolitanism: A Reconstruction. *University of Toronto Quarterly*, 77.

Wittgenstein, Ludwig. 2008. *Tractatus Logico-Philosophicus*, London: Routledge & Kegan Paul.

8

TAGORE–WITTGENSTEIN INTERFACE

The poet's activism and virtue ethics

Sitansu Sekhar Chakravarti[1]

At the ceremony for conferring the honorary D Lit on Tagore in 1940 by Oxford University, the spokesperson of the university designated the poet as 'myriad-minded' while introducing him to the audience gathered on the occasion thus:

> Here before you is the myriad-minded poet and writer, the musician famous in his art, the philosopher proven both in word and deed, the fervent upholder of learning and sound doctrine, the ardent defender of public liberties, one who by the sanctity of his life and character has won for himself the praise of the whole of mankind.[2]

The multifarious levels of human interaction, as indicated above, on top of the pursuit of the muse that Tagore would deeply involve himself in, often tend to baffle others, as the tie uniting these widely diverse areas is apparently found missing. A natural outcome of the puzzle is to emphasize some of these areas of involvement to the exclusion of the others, the latest example of which is found in a recent article by the Tagore scholar William Radice entitled 'What Makes a Writer Relevant?'[3] Here the author highlights the music, dance and creative writings of Tagore, and goes on to suggest that we need not pay as much attention to his other works and involvements which, to the author's mind, may not be of an abiding interest to posterity. 'Future generations,' he says, 'come along in the Golden Boat,[4] take his works away and make new, creative use of them, leaving Tagore the *Gurudev*[5] behind on the river-bank, for 'we don't need to go to

Tagore for ideas'; '[m]any people have those ideas already,' and '[h]is works – whether poems, songs, stories, novels, plays or paintings – were not just vehicles for ideas and ideals.' I fail to be convinced by the train of thoughts here. It is not only the case that without recourse to his ideas, and the ideal of life he held dearly on to, we are not able to grasp his creative works in an overall balanced way encompassing the various moods they portray; the ideas have a uniqueness of their own not to be found present in other thinkers unlike as Radice seems to claim. Tagore attempted his whole life to make the ideas clear to himself and others, an ongoing process indeed befitting their depth and breadth as well as the novelty involved in them. Indeed, I find it a worthwhile endeavour on the part of the Indian Institute of Advanced Study, Shimla, to arrange deliberations on Tagore's concept of reality where ideas from other great world thinkers may help shed light on the complex concept of reality itself. The secular spirituality constituting the texture of reality for him is certainly worth studying in connection with the concept of secular spirituality of Wittgenstein of the *Tractatus* days that mark a strong attachment of the philosopher to the poet. All this, needless to say, relates intimately to the phenomenological concept of reality that Tagore highlights in his conversation with Einstein, incorporating a very special kind of existentialism and encompassing the essence of the age-old Indian phenomenological approach. It is important to mention that the poet developed his own distinctive approach to existentialism at a time when the wave of the existentialistic movement had not taken possession of Europe yet. The follow-up of this of course is a unique kind of virtue ethics centering around the virtue *harmony within*, which grounds the Aristotelian virtue *sympathy*, and that to my mind deserves a special attention in today's academic world. The study of Tagore along with Wittgenstein may hopefully lead to an in-depth understanding of parts of the *Tractatus* ushering in a welcome moment for *Tractatus* studies.

Tagore and Einstein: the poet's phenomenological reality

The world, to Tagore, is a 'human world,' not in the sense of philosophical idealism, but in the phenomenological perspective, as reality reveals itself 'through art, literature and the religious consciousness of man'[6] as well as through science. The 'scientific view of [the world],' even though it deals with the 'impersonal,' 'is also that of the scientific man' as circumscribed by the methodology befitting the scientific mind, and thus ends up displaying 'the impersonal human world of truths.' Spiritual pursuit goes beyond the scientific in 'applying values to truth and we know truth as good through our own harmony with it,' signifying the existential moment of becoming. Thus, neither truth nor beauty is independent of the human being to Tagore. Granted there is an independent reality, it is accessed only in so far as it is grasped by the human mind in its phenomenological appearance, and the resulting truth 'therefore is human.' 'If there be some truth which has no sensuous or rational relation to the human mind it will ever remain as nothing so long as we remain human beings.' 'It is not difficult to imagine a mind to which the sequence of things happens not in space, but only in time like the sequence of notes in the music. For such a mind its conception of reality is akin to the musical reality in which Pythagorean geometry can have no meaning. There is the reality of the paper, infinitely different from the reality of the literature. For the kind of mind possessed by the moth, which eats the paper, literature is absolutely non-existent, yet for Man's mind literature has a greater value of truth than the paper itself.' 'What we call truth lies in the rational harmony between the subjective and objective aspects of reality. . . .'[7]

'Phenomenology', Mohanty says, 'asks us to focus on the way things are presented in consciousness, on the meanings that things have on those experiencing consciousness. Understanding consciousness as intentional and meaning-giving, phenomenology raises consciousness, in its transcendental (i.e., world-constituting) role, as the foundational principle for philosophy.'[8]

Wittgenstein and Tagore: I. The *Tractatus*

Things perhaps might become a little clearer if we see Tagore in proximity to Wittgenstein's *Tractatus*. The philosopher is known to have got intimately drawn to the poet's works during the early twenties of the last century, especially his play *The King of the Dark Chamber*, copies of which he took care to lend, or present periodically to his close friends and acquaintances. He read the book together with his student at Cambridge, Yorick Smythies, on more occasions than one, once around the time of his famous talk on aesthetics at the university in 1938. The book had cast such an abiding influence on his mind, that he took the initiative to 'translate' Scene II of the play, in company with his student, changing some expressions in the original translation by what the two considered 'idiomatic equivalents, . . . endowing the language with the immediacy they wished it to possess.'[9] To my mind, the attachment to Tagore's work in general, and specifically the play, had philosophical relevance for Wittgenstein, apart from its literary worth. In his book *Tractatus Logico-Philosophicus*, which had by that time started attracting attention of philosophers and mathematicians across the continents, Wittgenstein had already developed his views about science to the extent that it does not involve itself with value as object of investigation. Value indeed lies outside the realm of scientific pursuit. The *world* of Wittgenstein is bereft of value in so far as the world is there for the scientist to analyse toward attainment of truth. What *is* is the world with its constituents. About value Wittgenstein 'says':

> 6.41 The sense of the world must lie outside the world. In the world everything is as it is, and everything happens as it does happen. *In* it there is no value – and if there were, it would be of no value.
>
> If there is a value that is of value, it must lie outside all happening and being-so. . . . It must lie outside the world.[10]

However, since sayables can only be said in propositions that relate to the world, we cannot speak of the values that lie beyond. The world of Wittgenstein is the logical correlate of science, at the ontological level, which investigates its structure and happenings where the values are not included, while the human values limit the world as conceived by Wittgenstein in the *Tractatus*, reducing it to a 'limited whole,' and '[t]he feeling of the world as a limited whole is the mystical feeling' for the philosopher (6.45). One must note, Wittgenstein does not end up saying simply that the world is limited, for he considers it a limited *whole*, apparently complete in itself in respect of the parameters of science, and is limited nonetheless by the values falling beyond them, not spatially of course, but logically, – values that the world fails to accommodate by virtue of the logic of science. From what has been said it follows:

> 6.52 We feel that even if *all* scientific questions be answered, the problems of life have still not been touched.

As regards the solution of the problems of life, he had 'said' earlier in the *Tractatus*:

> 6.4312 . . . The solution of the riddle of life in space and time lies *outside* space and time.
>
> (It is not problems of natural science which have to be solved [in this process].)

For,

> 6.432 *How* the world is, is completely indifferent for what is higher. God does not reveal himself *in* the world.

The laws of science can be expressed in the descriptive language comprising Wittgenstein's propositions, which is not a vehicle for the values that are higher. These laws may hold their mystery for us at every step of scientific query and investigation, though without a trace of mysticism in the process, because there is no inbuilt transcendence to the values in the scientific pursuit, bracing the mystical, the inexpressible. The inexpressible here, however, is not a metaphysical entity, but relates to the values phenomenologically present, though not ontologically posited, which Wittgenstein finds *shown* in the play *The King of the Dark Chamber* and not described with use of propositions. No wonder the play fascinates him. The deep problems and concerns of life indeed relate to the mystical dimension which has a far more intimate touch with the human being than science does toward solution of life's problems. Tagore no doubt shares a lot with Wittgenstein here.

There is ample evidence that Tagore had his own criticism of the claim of science extending over virtually all areas of human interest. 'Our scientific world,' the poet says, 'is the world of reasoning. It has its greatness and uses and attractions. . . . But when it claims to have discovered the real worlds for us and laughs at the worlds of all simple-minded men, then we must say it is like a general grown intoxicated with his power, usurping the throne of his king. For the reality of the world belongs to the personality of man and not reasoning, which, useful and great though it be, is not the man himself.'[11] 'Science' he continues,

> sets up an impersonal and unalterable standard . . . which is not the standard of creation. Therefore at its fatal touch the reality of the world is so hopelessly disturbed that it vanishes in an abstraction where things become nothing at all. For the world is not atoms and molecules or radio-activity or other forces, the diamond is not carbon, and light is not vibration of ether. You cannot come to the reality of creation by contemplating it from the point of view of destruction. . . . [W]hen you deprive truth of its appearance, it loses the best part of its reality. For appearance is a personal relationship; it is for me . . . 'appearance,' which seems to be of the surface, but which carries the message of the inner spirit. . . .'[12] 'This

> [awareness] does not lead us to power as [scientific] knowledge does, but it gives us joy, which is the product of the union of kindred things.'[13]

'Thus, truth is not exhausted by the scientific pursuit of knowledge. Physical science tries to explain facts – not the relation that they bear in the domain of our values.'[14] The poet extols the value of mysticism falling beyond the jurisdiction of scientific methodology even though he may not feel prone to posit objects at the metaphysical realm as correlates to the mystic thrust. Hopefully we can see a lot of similarity between the two geniuses of our times in what follows, and have a new look at some of the passages of the *Tractatus* as well as the *Notebooks* to be able to suggest a new understanding, if not an alternative interpretation, drawing mainly upon the ideas contained in the play by the poet. I intend to point out some important differences in their positions as well, while we see the similarities. In the process I have taken care not to tie the discussion to Wittgenstein's picture theory of meaning, which he later abandoned while retaining a lot of the spirit of the *Tractatus*.

Before proceeding on, it might be advisable to highlight the distinctive ways that Wittgenstein and Tagore point to the limits of science. Both agree that values are outside its range, while Tagore goes further, questioning the claim that science deals with truth independent of human intervention, as he incorporates the phenomenological thrust pertaining to reality, which is reality for the human mind. As regards the question whether the stars are constantly moving, as suggested by science, and not standing still, as we see them at one single moment, Tagore claims that 'both facts are equally true about the stars.' 'The stars,' he says, 'in their one relation to me are truly still and in their other relation are truly moving.'[15] From 'the near view,' the earth is flat, while not so from the distant view. 'The distant and the near are keepers of two different sorts of facts, but they belong to one truth. . . . There is a point when in the mystery of existence contradictions meet; where movement is not all movement and stillness is not all stillness; where the idea and the form, the within and the without, are united; where infinite becomes finite, without losing its infinity. If this meeting is dissolved, then things become unreal.'[16] We wonder if Tagore gives us here an inkling of the truth contained in Quantum Physics pertaining to the *Superposition Principle*, i.e., of a probabilistic description of an object residing in multiple potential states, and its assuming or falling into a specific *Eigenstate* upon observation, even as the poet deals with things at the macroscopic level,– at a time, back in 1917, when the scientific theory was yet to be formulated.[17]

Wittgenstein and Tagore: II. *The King of the Dark Chamber*

In the play the King is invisible; the queen is both enchanted as well as fearful of him, having failed to form a convincing idea about him in a visible shape due to limitations that a conventional frame of reference suitable for the understanding of ordinary day to day living situations would dictate. The grandfather and his group, along with Surangama, the queen's maid, have, by contrast, a relation of love, hope and intimate trust in the King while they succeed in connecting to him in all situations of life. Here insistence on the traditional ways of interface and understanding are absent.

> My beloved is ever in my heart;
> That is why I see him everywhere,
> sings a group of minstrels.[18]

The King, in his turn, is very different from the other kings, for he does not force his ways on his kingdom ensuring law and order. 'Where would be the necessity of having a King if order and harmony existed already [without him]?,' says Bhavadatta, one of the foreigners who is totally puzzled by the apparent absence of a King in the country he is visiting and yet finds to his amazement 'order and regularity prevailing all over.'[19] The King 'exists' nevertheless, if not for the running of the kingdom of which a king is supposed to be in charge normally elsewhere, standing rather for the values we all live for, values that are there waiting ever to inspire us irrespective of the way we may feel inclined to respond to the King's intimately existential beckoning. The King personifies human values that limit the world as conceived by Wittgenstein in the *Tractatus*.

'The play is an allegory[20] of religious awakening,'[21] says Monk, 'and it echoes many of Wittgenstein's own thoughts on the subject.' The queen grows increasingly impatient to have a look at her husband, whom she has never seen, in a distinctly visible form, and keeps on imagining his beauty and grandeur all the time as she feels prone. At one point she refuses to continue meeting him in the dark chamber any more where nothing is visible, and comes out in the open to be with the 'king' imposter, to eventually realize that she has been deceived. The moment finally comes when her agitated mind calms down, while she readies herself to be in company of her dear King in terms appropriate for the occasion, without insisting on any imposed parameters that do not fit in. The King now happily consents to come out in the open with her, abandoning the dark chamber, as she is prepared to meet the world this time with him in the appropriate way. His invisibility ceases to be an issue when she has him out there with her in the world.

The invisibility of the King in the play indicates he cannot be grasped in worldly terms as standardized by science. Certainly, the values are not '*in* the world' (6.41) in this sense, and nothing can be said about them using language meant for saying things pertaining only to our ordinary involvements. He is not an entity in the world in as much as he stands for the values. Nor is he a metaphysical entity in so far as he is the poet's personification of the values in the play, more real as a phenomenological being in *the poet's* imagination than a metaphysical one, for the phenomenological pull is deep and persistent to the individual.

Tagore's King, however, is not absolutely outside of the world, i.e., the world of science in the Wittgensteinian sense, even though he evades us so long as we look for him in our exclusive terms, and the world continues to remain abstracted at our end. With the dimension of values added to it, the world is certainly richer as it stands compared with the restricted world of science. The King is not away from the world, for people can relate to him, and the world, as the song quoted above suggests, while they are *in* the world, transcending its narrow confines at the same time, being imbued with values. However, one has to go beyond the scientific parameters of knowledge to have an interface with him out in the open, in love and affection, scaling the confines of our narrow bounds that keep us away from harmony within and without. Prior to reaching this state our contact with him is only in the dark chamber where he is attractive from a distance as the beckoning calls of values reach us before we are ready to embrace them in their terms. In other words, the values are there in our unconscious mind, innate and a priori,[22] waiting to surface at the conscious level only when we are ready. The queen could hear the enchanting notes played on the King's musical instrument 'that called [her] through the lone and waiting night' only when she was willing to pay attention.[23] One can respond to 'his call any way' one likes, for he is prepared to arrange ways of welcome to suit the various responses to the inner calls.[24] His call is silent, so is his coming. At long last when the queen is ready to receive the King outside of the terms of her imposition, he joins her in her dealings with the world. Certainly her world 'waxe[s] . . . as a whole' (6.43) with the values incorporated. The King, we must note, does not impose himself on the queen; he is not shown to physically rescue her when the palace is set on fire by a king from another country, who harbours a hidden desire for her in the physical absence of the King. The King does not abandon her either, for he waits for her to find the way to life existentially herself, ushering in the moment his presence would be truly relevant.

The King is God, though not an entity in the world, nor a metaphysical entity outside, and poses a limit to the world of science in respect of the values he personifies. According to Wittgenstein, such a phenomenological awareness 'change[s] the limits of the world, not the facts; nor the things that can be expressed in language,' (6.43) as Wittgenstein 'says' in the *Tractatus*. The following lines in continuation seem quite relevant:

> In brief, the world must thereby become quite another. It must so to speak wax or wane as a whole. The world of the happy is quite another than that of the unhappy (6.43).

The King in the play does not rule the world, nor mete out punishments, even though Surangama, the queen's maid, takes him to have 'sent away [her] father in exile' for his drinking and gambling spree.[25] Things happen in the world according to the laws of nature, and moral lapses have their own results. The world of Tagore, however, has the value of goodness imbedded in it, if we have eyes for it, transcending the scientific level, so that co-operating with it in the process of existential becoming helps us establish harmony within and around. Love, that keeps the process alive, is not an isolated, unreal element at the micro level of the individual, but connects us, the individuals, to the macro level of the universe as well, so that there is a continuity all the way, emanating from the macro, while existentially accommodated at the micro. This I believe is what the queen means when she says to the King at the end of the play:

> Your love lives in me – you are mirrored in that love, and you see your face reflected in me: nothing of this [is] mine, it is all yours, O lord!

It might be a little far-fetched to see here Tagore involved in the metaphysical relation between the creator and the individual, as Monk's discussion seems to indicate. The world with values is the *total* 'human world' for Tagore that incorporates the world of the 'scientific man' and is phenomenologically present to human consciousness.

The Wittgensteinian world in the *Tractatus*, in contrast, is absolutely severed from values; the transition to the values takes effect from this secular world of science without effecting any change there, as we have already noted. In a way the queen also starts her journey from a world that is secular to a large extent, though not absolutely so, for she receives the calls *in* there even when she is not ready for them. However, with her phenomenological/existential transformation this world gets transformed for her, so that she can live and relate to it in harmony, with the values added. Sudarshana, the queen, is not just a transcendental subject. For her, it is not that she is adding something foreign to the world which does not belong here; she is simply able existentially to attune herself to the already existent. For Wittgenstein, speaking of values results in the issuing of nonsensical sentences. Tagore does not go by the Wittgensteinian picture theory of meaning, so he can speak of the world with the values in the poetic mode in his play. The *poet* has always occupied a special place in Tagore's writings, because it is the *poet* who can say things meaningfully, according to him, as is evident in another play by him, *The Cycle of Spring*. The 'contemplation of the world sub specie aeterni' does not limit Tagore to 'its contemplation [only] as a limited whole' (6.45). It invites him to take part in the world's activities, even in an activist way, as he did throughout his life. Here he parts ways with Wittgenstein.

The state of redemption through acceptance of the values incorporates the unique strain of Tagore's existentialism that accommodates involvement in the world in the right spirit, without resigning to a detached mood of awe and reverence. To Tagore, an abiding goodness is the telos of the universe, which inspires us to move in the desired direction, making it our own goal. In the process of evolution, human beings acquired their sense of value which they have an existential choice to make their own. Tagore dreamt of a next step in the process where human society would rise to a new height, established in the virtue *harmony within*, in existential co-operation with this telos:

> Here rises the great humanity,
> Shivers go around . . .
> The moment of the great birth has indeed arrived . . .
> Victory be to the ascent of humanity.[26]

It is not unlikely the play appeared to Wittgenstein as indicating, though in ways he would designate as nonsense, the right philosophy of religion, coming very close to his own, if we are permitted to say, where things about God are not said, but indicated or 'shown' in the poetic way.

The secular spirituality of Wittgenstein may perhaps be looked upon as his vision of Christianity without belief in God as an entity mundane or metaphysical. We must keep in mind that although Wittgenstein does not posit God as an entity, but only as a value in the *Tractatus*, he has an abiding respect for people's reverence for God, in marked contrast to Freud and Marx.[27] The value certainly does not relate only to the totality of the world (cf., 6.43 above), ignoring the constituent facts, nor can there be a resulting sense of harmony in the world accompanying the perspective of 'sub specie aeterni' (6.45) when the constituents get ignored.

In the *Notebooks* where Wittgenstein says:

> How things stand is God.
> God is how things stand. (1.8.16),

he does not need to be taken as going against the *Tractatus*, which holds that value must lie outside of the world. Suggesting that the totality of facts, i.e., the world as a whole, is the holding ground for the value[28] may not help, because one may not accept the totality as being *outside* of the world. True, in the *Tractatus* one has to go beyond the world to look for value. However, once one has reached the mystic state having gone beyond, looking back at the world is never the same again. With the mystic transformation having taken shape in the subject, the facts (i.e., how things stand) have got a touch of value on them. According to Wittgenstein, it is possible for one to phenomenologically transcend the world without actually bringing value into it, although to such a person, after transcendence has been achieved, the world does *appear to be* touched with value, so that the world 'wax[es] . . . as a whole' only phenomenologically, with the facts taken both jointly and severally, while the individual merits of the facts, or lack thereof, cease to be emphasized in isolation (TR 6.43) as the world is viewed at this point 'sub specie aeterni' (6.45).[29] Such a world *viewed* with the touch of value is God in the *Notebooks*, not a world incorporating the value (1.8.16). Thus, in the *Tractatus* and the *Notebooks* we find two points of view regarding the relation between the value and the world, one from the bottom up, that is, from the world to the value, as in the *Tractatus* (6.41, 6.432), and the other just the opposite (TL 6.43, *Notebooks* 1.8.16); they are complementary to each other, when taken in the right perspective, and certainly not contradictory.[30]

All this reveals to us why it would be normal for Wittgenstein to be philosophically drawn to *The King of the Dark Chamber* by Tagore, while we find some passages in the *Notebooks* as well as the *Tractatus* making sense together in the light of the ideas contained in the play that fascinated the philosopher. To repeat, the world is totally different to the queen before and after her surrender to the value that the King personifies, even though it has not undergone any real change around her; the difference lies only in how she views and deals with it, having established harmony manifest phenomenologically within and without.

The spirituality of Tagore is secular in its own way. When the poet composes songs on the Hindu deities, he by no means needs to be taken as believing in their existence abandoning his Brāhmo stand of a formless God. He is committed here to the deep meanings relating to the icons, not their ontological reality. He speaks a lot on the subject of spirituality, without making any linguistic reservations as Wittgenstein does, and addresses God in his numerous poems and songs, as well as portrays Him in his plays. His spiritual secularism is not via linguistic analysis as in Wittgenstein; it is phenomenological. It is quite normal to question if he is, after all, committed to a personal God, or an impersonal entity for that matter, when he addresses Him in his creative writings. Tagore certainly seems to have no cosmology other than that provided by science. Thus, we mark the presence of a secularism of ontological non-commitment at the root of the poet's mystic mood grounded on phenomenology that accommodates, and does not exclude. It is interesting to note in passing that Tagore wrote a short story, 'The Hungry Stones,' where a mystic atmosphere of the phantom world is created without a commitment to the existence of phantoms. The parallel situation for Wittgenstein is that when he is involved in doing analytical philosophy in the *Tractatus*, he is phenomenologically open to the values even as he finds it logically impossible to put them into words.[31]

Harmony and Tagore's virtue ethics: interrelation of ethical and aesthetic

Clarification on *harmony* is overdue. Let me make up for the lapse now, for this is essential for understanding the line of thinking, leading on to Tagore's virtue ethics on which I would like to say a few words as a follow-up to his concept of reality. The expression 'harmony' has two senses: (1) the first refers to a harmonious state that obtains on its own out there, or is brought about on the basis of actions undertaken. The harmony one finds present in nature is of the former kind. Human intervention toward mass exploitation of nature eats into it and ecological imbalance results. Those who believe in Gandhian socialism think that the societal harmony it aims at is to be brought about by human involvement the right way. (2) The other sense pertains to an attitude, to wit, a disposition (courage being a mental disposition), typically a character state, which again one can work on toward its furtherance, as one can on courage. The Gandhian socialism is dependent on building this harmonious state within, on the basis of which one would be motivated toward proper actions leading to the goal of building a harmonious society. Harmony within is a disposition that inspires to build harmony around, resulting in doing good to others as well as contributing to ecological harmony. Here harmony is not just balance or equilibrium; like harmony in music, it is creative in its aesthetic mode present in the ethical dimension, and is pleasing overall – certainly not binding or depriving, either to oneself or the other. Thus, harmony is not simply self-control that Aristotle talks about as a measure against self-indulgence or intemperance, but adds a meaning to the onward flow of life. Harmony in this second sense is a virtue which we prefer to designate as *harmony within*. To sum up the ideas above, harmony has a threefold aspect: (a) the *harmony within*, which is the virtue; (b) the harmony amongst human beings – in one's own society and the whole social world – that the virtue is meant to, and helps promote; and (c) the harmony of the individual and every society with the natural world, which also the humans look for and delight in promoting on the basis of the virtue.[32]

The virtue *harmony within* is a character state, as indicated above, that equips, and prompts, the possessor of the virtue to furtherance of harmony within and without, a process accompanied by a sustaining, unconditional joy manifest even in the midst of failures and many challenges on the way, an intrinsic value which the person intends to share with others. The process accommodates pluralism while not subscribing to moral relativism. Pluralism is present here in an existential setting in so far as there is possibility of a scope for people to react differently to situations depending on the state of harmony an individual is in within the frame of reference of the specific orientation of her nature, taking into consideration the cultural diversity in place. However, pluralism is nested here in the concept of a general, overall harmony where the good of all is an end encompassing a holistic outlook of well-being that does not isolate nature from its fold, and is not measured only in terms of material attainment. To me it seems Tagore subscribes to the virtue *harmony within*, as explicated above, in his views on ethics. He says:

> The methods of our moral training have been based upon the fact that by changing our mental focus, our perspective, the whole world is changed and becomes in certain respects a

> different creation with things of changed value. Therefore what is valuable to a man who is bad becomes worse than valueless when he is good . . . rearranging the meaning of things in different proportions and forms. Such mobility of mind plays havoc with things whose foundations lie in fixed convention.[33] . . . Institutions which are so squarely built, so solid and thick, become like vapour in th[e] poet's world. . . . On the other hand, love of comrades, which is a fluid thing in the ordinary world, which seems like clouds that pass and repass the sky without leaving a trace of a track, is to the poet's world more stable than all institutions. . . . He perceives in his world that this love of comrades, like clouds that require no solid foundation, is stable and true, is established without edifices, rules, trustees or arguments. . . . When the mind of a person . . . moves in a time different from that of others, his world does not necessarily come to dislocation, because there in the centre of his world dwells his own personality. All the facts and shapes of this world are related to this central creative power, therefore they become interrelated spontaneously. His world has its own consistency . . . because of the central personal force.[34]

It is the Tagorean sense of harmony, which is an aesthetic concept that binds all his involvements together in a broader aesthetic mode, where ethics finds its place. Here he may not toe Wittgenstein's saying:

> 6.421 Ethics and Aesthetics are one and the same.

To Tagore, *harmony*, in the aesthetic mode, is the basic virtue for us humans. It goes deeper than empathy, which by itself may not be enough in a situation for justified moral acts to follow, for the latter have to be backed by proper understanding of the relevant situation, and the required expertise relating to it, on which a right act can stand. The virtue *harmony within* encompasses empathy and incorporates within itself the dimension of right understanding and the needed expertise. *Harmony* allows us both to honour as well as promote a value.[35] Here we not only have an advantage over Michael Slote's care ethics, but also over consequentialism as highlighted by Pettit in so far as the position we are depicting does not accommodate a mechanical decision procedural technique, leaving room for honouring the value.

Living itself is an aesthetic pursuit to Tagore, where the ethical focus is contained as part of this larger pursuit. An action is wrong in so far as it hinders the harmony of life. However, even though ethics is embedded in aesthetics, it has its own logic as rooted in the virtue *harmony within*, and is not reducible to the other. Aesthetic pursuit, in its turn, is circumscribed by ethical consideration for Tagore. Exercise of the virtue *harmony within* toward establishing or maintaining harmony for the welfare of the animate and the inanimate world is a necessary condition for a moral act, and often a sufficient condition as well. In the creation of art or literature, however, the prime focus is not establishing or maintaining harmony but the aesthetic expression of harmony itself, although while developing the expression the sensitivity of the aesthetic creator may not allow her to abstract or estrange herself from the truth of harmony in all beings which manifests itself in the good of all, thus leading her to the task of establishing it when found lacking. It is in this context, I believe, that '[a] true artist is also an ascetic' according to Tagore.[36] The moment of aesthetic creativity, in its turn, however, does not involve itself in performing an act in the sense an activist does. The creator of beauty is not a preacher or a person of action. However, she does relate attitudinally to others no less on this count as she is grounded on the truth of life. The abiding joy on the face of truth is realized by the poet, and 'expressed in poetry and literature.'[37] Although the scientist is after depicting truth, finding, or giving expression to, the joy on its face is never his task. No wonder, *the poet* as a character in Tagore's plays exerts such a pronounced influence on the other characters. Tagore's own songs have inspired the freedom fighters in India's struggle for independence, including Mahatma Gandhi, as well as the people of Bangladesh in their rise against repression. Needless to say, Tagore has often involved himself, in his own ways, in the role of an activist for the cause of humanity.

Ethical grounding in harmony as such is, by itself, we may say, a far cry from harmony of the full-blown creativity of the aesthetic kind of the poet or the artist. A dance performance, or recitation of poetry, is hardly ever considered an example of a moral act on its own. Likewise, an exemplary moral act per se is not automatically deemed as incorporating high aesthetic value. Ethical personalities, such as Gandhi, may lack in a special urge for the joy of aesthetic creativity, not even through enjoyment of others' creation, while they remain steeped in the virtue *harmony within*, and carry on their activist programmes with a zeal toward establishing harmony among humans, in tune with the physical environment. They strive for an end to exploitation, human or ecological, that generates disharmony resulting in suffering. Radice's reference to Tagore's pining for his aesthetic creativity while immersed in his activism 'harnessed to a cause,' with 'a hundred good intentions,' simply goes to highlight that ethics is not a substitute for aesthetics even though they are closely connected in that the former is grounded in the latter in a broad sense. Tagore, to my mind, never alludes to an inbuilt antagonism existing between the two as Radice seems to suggest. There, thus, is no need on our part to advocate Tagore's aesthetic pursuits only, as Radice does, to the exclusion of his other involvements unless we are ready to have our Tagore abstracted, contrary to his basic holistic thrust.

Tagore's difference with Wittgenstein comes out clear at this point. The former is a creative artist, poet, singer and choreographer on the one hand, while an environmental and social activist on the other in as much as aesthetics and ethics are intertwined; whereas for Wittgenstein ethics may be seen as reduced to the aesthetics of wonder at the 'view' of 'the world sub specie aeterni' (6.45) where no action is called for on the part of the agent. We must keep in mind Wittgenstein was quite critical of Russell's activism.[38]

Conclusion

Tagorean spirituality, centred around the concept of harmony, is embedded in his understanding of the Indic tradition. Harmony is the tie uniting the Indian Tagore with the global. However, we can locate the poet's spirituality centering around harmony only with a proper understanding of his philosophy, a task that seems impossible to perform with grounding only in traditional Indian philosophy. Tagore was a universal man. The thoughts he tried so hard to articulate both in India, mainly in Bengali, as well as in the West, in English, and might not have succeeded all the way since language appropriate enough to express their novelty had yet to emerge, can be accessed today by combining the Indian way with the modern Western. The task has not begun yet to explicate the *rasa-yoga* (i.e., the yoga of aesthetic creativity)[39] of Tagore, which encompasses the *yoga*s of knowledge, devotion, action, as well as service toward amelioration of suffering, crystallizing his contribution to the Indian tradition, and the Western.

Tagore repeatedly denounced the claim to his being a philosopher. In a way he was justified in the disclaimer in that he never did philosophy in the traditional way. He seems to have some similarity here with Wittgenstein. Neither of them usually makes references to others' works in order to establish their own philosophical positions. As in the case of Wittgenstein, Tagore adopted his own way of philosophizing without taking primarily to the way of argumentation. His novelty consisted in incorporating the poetic language and mood in his philosophical reasoning. I am not speaking here of the application of his philosophical insights to his creative writings, as in *Post Office*, *Red Oleanders*, or *The Cycle of Spring*. Jean-Paul Sartre leaves the mark of his philosophical thoughts in his creative writings, while feeling the need to pursue philosophizing independently. So does Tagore, although the method followed by Tagore in the latter pursuit was a new, holistic way of reasoning, where he combines reason with poetry. In the case of Wittgenstein, his contemporaries have tried to clarify his philosophical positions applying the traditional methodology. This, to my knowledge, has not been attempted for the philosophical thoughts of Tagore yet. I believe the phenomenological approach would be quite appropriate here. This will not only open up a new way of seeing things in the world today but also provide a guide to a new direction of involvement with the animate and inanimate worlds. In this process Wittgenstein's mysticism will hopefully find a much needed understanding that it deserves.

We come to the end of a full circle here. I started the chapter with mention of the multi-faceted involvements of Tagore for which the Tagore scholars might fail to find a unifying tie. First, I dealt briefly with the topic of Tagore's concept of reality as it transpires in his dialogue with Einstein. For a deeper understanding of the issue I juxtaposed the poet's stand with that of Wittgenstein in the *Tractatus* focusing on the poet's play *The King of the Dark Chamber*. This led to an explication of the multi-faceted concept *harmony* that goes to the heart of his ethics as embedded in aesthetics, grounded as the latter is at its base on his concept of reality. Pairing Wittgenstein's dealing with the nature of reality that spills onto the areas ethics and aesthetics in his *Tractatus*, I have brought to bear the deliberations on Tagore's concept of reality to his consideration of these two areas. The starting point of our attempts at making sense of Tagore's concept of reality as appearing in his discussion with Einstein hopefully blossoms out into a meaningful explication of a theory as the poet's view is placed in the philosophical context of Wittgenstein's *Tractatus*.

Notes

1 I must thank Yves Roberge, Principal, and June Larkin, Vice-Principal, New College, University of Toronto, for providing all possible help toward preparation of this chapter. I am indebted to Shankha Ghose, the foremost Tagore scholar of the day, for his comments on an earlier version. The long correspondence I had with Michael Hodges on Wittgenstein's mysticism was of great help. While writing the chapter I bore in mind the observations and encouragement by Rosalind Hursthouse on my initial account

of the Indian version of Virtue Ethics centred around the virtue *harmony within*. My son Ananda Roop's constant participation in the preparation of the chapter, both with technical help on the computer as well as sharing his views with me on its content, merits a mention. I am thankful to Thomas Loree for making time in the midst of his tight schedule to copy-edit a previous draft of the chapter on which the final version has been made. I am painfully aware of the fact that several points dealt with here deserve more elaborate treatment than space and time would permit.

2 Robinson and Krishna, *Rabindranath Tagore: The Myriad-minded Man*, p. 353.

3 William Radice, 'What Makes a Writer Relevant?' *The Hindu*, Chennai, May 7, 2011.

4 Here Radice is referring to the poem by Tagore of the same name, where the poet is dismayed that the carrier is filled to the brim with the harvest of the poet's own creation, without the barest room left for him.

5 The expression means *the revered teacher*.

6 Rabindranath Tagore, *The Religion of Man*, Appendix II ('Note on the Nature of Reality'). The rest of the quotes in this section are from the same appendix unless indicated otherwise.

7 See endnote 6.

8 J. N. Mohanty, *Between Two Worlds, East and West: An Autobiography*, pp. 111–12.

9 Ludwig Wittgenstein, 'The King of the Dark Chamber', *Wittgenstein's Whewell's Court Lectures*, Chapter 11, p. 325.

10 Max Black unfortunately takes these words literally. (c.f., Black (1964), p. 370.) Wittgenstein certainly does not mean that there is space *outside* of the world. What Wittgenstein seems to mean is that one is phenomenologically aware of the values, which cannot be accommodated in the world as it is conceived. Thus, the values pose a limit (not a spatial limit) to the world reducing it to a 'limited whole'.

11 Tagore, *Personality*, p. 52.

12 Ibid., pp. 50–1.

13 Tagore, *Sadhana*, p. 21.

14 Sitansu Sekhar Chakravarti, 'The Spirituality of Rabindranath Tagore', p. 271.

15 Tagore, *Personality*, pp. 43–4.

16 Ibid., p. 44.

17 I am thankful to my brother-in-law Amitabha Bagchi, a physicist, who helped me formulate this sentence so that it is accessible to physicists, without making any ungrounded claim about Tagore.

18 Rabindranath Tagore, *The King of the Dark Chamber*, p. 21.

19 Ibid., p. 19.

20 In a letter to C F Andrews the poet objects to categorizing the play as allegorical. 'Critics and detectives are naturally suspicious,' he says. 'They scent allegories and bombs where there are no such abominations. It is difficult to convince them of our innocence . . . the human soul has its inner drama which is just the same as anything else that concerns Man, and Sudarshana is not more an abstraction than Lady Macbeth, who might be described as an allegory representing the criminal ambition in man's

nature.' (Rabindranath Tagore, *Letters to a Friend*, New Delhi: Rupa, 2002, pp. 48–9.)

21 I feel inclined to say instead: *an allegory of waking up to the values.*

22 Ethical truths pertaining to *harmony within* and the way we relate to the virtue are contingent apriori, to use the Kripkean concept.

23 Tagore, *The King of the Dark Chamber*, pp. 187–8.

24 Ibid., p. 164.

25 Ibid., p. 42.

26 Rabindranath Tagore, *'Oi Mahamanaba Aashe' Geeta-bitan*, Visva-Bharati, Kolkata, p. 867. Translation of the passage cited is by the author of this article. Sandler develops on the ideas of Hursthouse relating to 'the ends constitutive of human flourishing', and suggests 'three promising candidates for components of human flourishing that are not appropriate for nonrational . . . social animals: meaningfulness, knowledge and autonomy.' (Ronald Sandler, *Character and Environment: A Virtue-Oriented Approach to Environmental Ethics*, 2007, p. 23.) To Tagore the end is harmony based on *harmony within*, nested in the aesthetic dimension.

27 The *Tractatus* ends with:

> What we cannot speak about we must pass over in silence. (7)

At the end of Wittgenstein's famous 'A Lecture on Ethics' the following lines are contained:

> . . . a tendency of the human mind which I cannot help respecting deeply and I would not for my life ridicule it.

28 This is Michael Hodges' suggestion.

29 Certainly, there is a change in the subject here *in* the world, which Hodges refers to in personal communication, pointing out an anomaly in Wittgenstein's position. Tagore's position, however, is free from the problem.

30 This is the most that can be made of 1.8.16 in the context of the *Tractatus*, which apparently contains the ultimate view of the philosopher on the subject. Here we do have a clue as to why *The King of the Dark Chamber* was so fascinating to Wittgenstein. I am thankful to Michael Hodges for drawing my attention to some passages in the *Notebooks* in his email correspondence with me in support of his own view regarding the totality being the reservoir of the value, to which I responded with my own observations as outlined above.

31 I have dealt at some length on Tagore's secular spirituality in my unpublished article 'The Tie Uniting Global and Indian Tagore: An East-West Philosophical Perspective.' The Tagore scholar Shashi-Bhushan Dasgupta, apparently corroborates the interpretation in suggesting that humanity appears to have taken the place of God to the poet with the passage of time. (See my article in Bengali: 'Wittgenstein's Mysticism in Light of Tagore's Thoughts', in Sudhir Chakrabarti (ed), *Rabindranath*, vol. 2, Kolkata: Institute of Human Development, 2011.)

32 I gratefully insert the words of Hursthouse, which she used to summarise my position in her email correspondence.

33 Tagore, *Personality*, p. 48.

34 Ibid., pp. 49–50.

35 Pettit, Philip et al., *Three Methods of Ethics: A Debate*, Malden, MA: Blackwell, 1997.
36 Rabindranath Tagore, 'Sense of Beauty', in *Angel of Surplus*, p. 54.
37 Rabindranath Tagore, 'Saundarya-bodh', in *Sahitya, Collected [Bengali] Works of Tagore*, vol. 8, p. 371.
38 Ray Monk, *Ludwig Wittgenstein; The Duty of Genius*, p. 211. Monk alludes to the incompatibility between the two philosophers pertaining to their 'difference between trying to improve the world, and seeking only to improve oneself.'
39 It accommodates all three traditional yogas, viz., knowledge (jnana), devotion (bhakti) and karma (action), within its fold, and yet is not exhausted by them.

References

Black, Max. 1964. *A companion to Wittgenstein's Tractatus*, Cambridge: Cambridge University Press.

Chakrabarti, Sudhir (ed). 2011. *Rabindranath: Bakpati – Bishwamana*, vol. 2, Kolkata: Institute of Human Development.

Chakravarti, Sitansu S. 2016. *Phalguni Nataker Antarer Katha*. ['The Inner Message of the Play Phalguni], Kolkata: Ebang Mushayera, vol. 23(2).

———. 2011. Rabindra Aloke Wittgensteiner Maramiya Tattva. In Sudhir Chakrabarti (ed), *Rabindranath: Bakpati – Bishwamana*, vol. 2, Kolkata: Institute of Human Development.

———. 2009. Ethical Message of the Mahabharata in the Wake of the Global Financial Crisis. *Journal of Human Values*, 15(2): 97–105.

———. 2006. *Ethics in the Mahabharata: A Philosophical Inquiry for Today*, New Delhi: Munshiram Manoharlal.

———. 1997. The Spirituality of Rabindranath Tagore. In K. R. Sundararajan and Bithika Mukerji (eds), *Hindu Spirituality: Postclassical and Modern*, New York: Crossroad.

Dasgupta, Shashi-Bhushan. 1992. *Upanishader Patabhumikay Rabindramanas*, Kolkata: Supreme.

Hodges, Michael P. 1990. *Transcendence and Wittgenstein's Tractatus*, Philadelphia: Temple University Press.

Hursthouse, Rosalind. 1999. *On Virtue Ethics*, Oxford: Oxford University Press.

Javidtash, Tomaj. 2015. *Nondual Perspectives on Quantum Physics*, North Charleston: CreateSpace.

Mohanty, J. N. 2002. *Between Two Worlds, East and West: An Autobiography*, New Delhi and New York: Oxford University Press.

Monk, Ray. 1990. *Ludwig Wittgenstein: The Duty of Genius*, New York: Free Press.

Pettit, Philip et al. 1997. *Three Methods of Ethics: A Debate*, Malden, MA: Blackwell.

Radice, William. 2011. What Makes a Writer Relevant? *The Hindu*, Chennai, May 7, accessed at www.thehindu.com/opinion/op-ed/article1996887.ece
Robinson, Andrew and Krishna Dutta. 1995. *Rabindranath Tagore: The Myriad-minded Man*, London: Bloomsbury.
Sandler, Ronald L. 2007. *Character and Environment: A Virtue-Oriented Approach to Environmental Ethics*, New York: Columbia University Press.
Slote, Michael A. 2007. *The Ethics of Care and Empathy*, London: Routledge.
Tagore, Rabindranath. 2000. *The Religion of Man*, Kolkata: Visva-Bharati.
———. 1980. *Personality*, New Delhi: Macmillan Company.
———. 1979. *Sadhana: The Realisation of Life*, New Delhi: Macmillan Company.
———. 1978. *Angel of Surplus*, ed. Sisirkumar Ghose, Kolkata: Visva-Bharati.
———. 1961. *The King of the Dark Chamber*, London: Macmillan Company.
———. 1960. *Geeta-bitan*, Kolkata: Visva-Bharati.
———. 1950. Post Office. In *Collected Poems and Plays of Rabindranath Tagore*, London: Macmillan Company.
———. 1928. *Letters to a Friend*, ed. C F Andrews, London: G Allen and Unwin.
———. 1925. *Red Oleanders*, London: Macmillan Company.
———. 1917. *The Cycle of Spring*, London: Macmillan Company.
———. 1907. *Rabindra-Rachanavali* (Collected [Bengali] works of Tagore), vol. 8, Kolkata: Visva-Bharati, (saka).
Wittgenstein, Ludwig. 2017. The King of the Dark Chamber, trans. in part by Ludwig Wittgenstein along with Yorick Smythies. In Volker A. Munz and Bernhard Ritter (eds), *Wittgenstein's Whewell's Court Lectures, Cambridge, 1938–1941, From the Notes by Yorick Smythies*, Malden: John Wiley, Chapter 11, pp. 325–33.
———. 2014. *Tractatus Logico-Philosophicus*, trans. Frank Ramsey and C. K. Ogden, Peterborough: Broadview Press.
———. 1979. *Notebooks 1914–1916*, London: Blackwell.
———. 1961. *Tractatus Logico-Philosophicus*, trans. D. F. Pears and B. F. McGuinness, London: Routledge and Kegan Paul.

9

TWO CONCEPTIONS OF REALITY

Tagore and Einstein

Syed Sayeed

1

More often than not when fundamental issues become the topic of conversation between a poet and a scientist, the result is a total non-conversation, a sterile exchange of selective ignorance (and often a display of bad attitude into the bargain), in which neither party shows itself in a creditable light. However, when such a conversation happens to take place between two great and generous minds, who are outstanding figures in their respective fields, there still occurs a miscommunication, but in this case it would be a rich and fecund miscommunication representing a site of cross-fertilization of ideas, a space in which the creative friction between two perspectives generates valuable insights.

We find this happening in the famous conversation between Tagore and Einstein when they met at the latter's house in July 1930 and the conversation turned to the question of the nature of reality. It is a happy coincidence that there seem to have been present people on whom the significance of this exchange was not lost, who took the trouble to record it and ensured that it was published soon after. Both the renown of the participants of the dialogue as well as the importance of the topic has led to innumerable comments and discussions, some of which are themselves extremely edifying. But repeated attempts even by mediocre minds at understanding great debates should not be scorned. They have their value. If nothing else they prevent us from forgetting the important questions that each generation must understand and answer in its own terms. The following is a modest attempt purely at that level. The reader, therefore, is advised not to expect any original arguments or new ideas. The effort here is just to make explicit what is implied by the views expressed by Einstein and Tagore by placing them in their properly philosophical context. Since this piece does not pretend to be a scholarly piece of writing, I have taken the liberty to omit detailed references to sources and so on.

2

Broadly, the task undertaken here consists of translating what is said by both the interlocutors into more explicit philosophical terms and place the two views in a common and coherent frame such that their ramifications can be fully understood. With Einstein's side of it, it is not so difficult to translate what he is saying into more precise (that is to only say easier to argue about) philosophical terms, partly because he is speaking from within the paradigm in which this issue had been extensively debated in modern times. But with Rabindranath's utterances, it is somewhat tricky. He does not couch his position in the prevailing philosophical idiom. One could imagine that this might be at least partly because of his being a creative writer who was not used to what you might call professional philosophical debates, and that the prevailing philosophical idiom did not come to him so naturally. But that looks unlikely. He had, by that time, long experience of serious engagement with philosophical questions and was by no means a stranger to the ways in which contemporary philosophers argued. In any case, it would be stupid to imagine that he could not muster a philosophical mode of statement and argument adequate for the occasion. Many lesser minds show enough aptitude for a facile mastery of that sort (and routinely make fools of themselves, but that is another matter). What one notices rather is a conscious refusal to abandon what looks like a mystical, poetic framework, for a discursive mode putatively more suitable for the topic on hand. It is impossible that he did not know that his choice of terms and concepts and his reference to 'superpersonal man' etc. are liable to make him sound not just mystical – in the pejorative sense – but hazy and woolly headed even to his immediate audience. Yet, he chooses to persist with his 'poetic-mystical' style of assertion. Evidently, he thought it important to stick to that terminology and deploy those very concepts. This requires that we exercise greater caution while translating what he was saying into more conventional philosophical terms. We must take care not to distort or dilute the core of his philosophical position in the process of translating his idiom.

The other part of the task undertaken here consists of placing both views in a common frame in order to make the structure of the debate more accessible. It must be pointed out that this is, strictly speaking, an impossible task since most if not all significant philosophical debates are necessarily trans-paradigmatic and a certain amount of incommensurability invariably attends them. However, we have no option but to accept recourse to it, always remembering that what we are doing can be justified only as a heuristic exercise wherein exactitude has to be sacrificed for the sake of facilitating understanding.

In the case of the dialogue we are discussing, its trans-paradigmatic character is more salient than in most cases, due to the fundamentally different philosophical positions from which the two interlocutors are speaking, as well as due to the fact that their views come from different civilizational perspectives.

There is another point we need to take note of. There are certain concepts which both of them seem to use loosely and sometimes in what looks like a significantly misleading way. At one level this is simply a consequence of the fact that they were having a cultured conversation and not debating in an academic seminar, and the use of technical jargon and an insistence on detailed arguments would have made the interaction tediously pedantic. At the same time, we must be wary of taking the casualness of usage at face value. At another level, it is like a short-hand notation of shared assumptions and accepted arguments. This means that if we wish to understand their respective positions and the exact points of disagreement between them, we must calibrate some of the expressions they use towards greater precision. There are quite a few terms which belong to this category. But I would like to draw attention to one concept which is very central and there is a danger of our going astray if we do not tread cautiously around that concept. It is the term 'truth.'

3

Generally, aside from the myriad problems that surround this concept, additional complications come from the two different ways in which the term 'truth' is used. First, is the use of truth as standing for some Platonic essence, often spelt with a capital T. The other is the use of the word truth to denote both truth and reality without further contextual clarification. As for the first source of confusion, it is the result of what you might call 'the mischief of nouns,' which analytic philosophers have talked about extensively. The origin of this confusion lies in the fact that verbs and adjectives often have their corresponding noun forms, and there is a constant temptation to look for some concrete or abstract entity answering to that term. Confusions regarding things like truth, beauty, value etc. are mainly owing to the fact that the form of the word leads us to assume a corresponding entity, and when it is not found in the ordinary way, a great deal of metaphysical ingenuity is exercised to locate and describe the elusive entity. One solution, as some sensible philosophers have suggested, is to ban the use of these noun forms. People are much less liable to get into confusion if they confine themselves to 'true,' 'beautiful' and 'valuable,' and not use the terms 'truth,' 'beauty' and 'value,' thereby reminding themselves that they are talking about attributes or relations. However, experience shows that linguistic recommendations from the viewpoint of philosophical clarity do not work. Linguistic habits die hard. (It is perhaps as well. Analytic philosophers would go out of business if they didn't.)

In the Einstein-Tagore debate, however, the source of the problem lies in the fact that both of them seem to use the term 'truth' to refer to both truth and reality at different points without an explicit caveat. As we will see, this is not just carelessness on the part of the interlocutors. There is a philosophical reason, though that is itself a matter of considerable debate. We shall glance at this aspect in passing later on.

4

Anyway, now that we have glanced at what we need to keep in mind while listening to this dialogue, we can move on to asking as to the main bone of contention.

The central, substantive issue is this: how coherent is the notion of a universe as it exists in itself, by itself, without any reference to the human Being, and if it is coherent, what are the ramifications of such a position.

Very briefly, Einstein seems to be saying that such a position is coherent and that it is on the whole unproblematically consistent with the way we think, feel and live. Rabindranath thinks otherwise.

The real problem here, as I hinted at in the beginning, is that this is not a straightforward confrontation of two different positions within a broadly shared framework. Quite early on in the dialogue Einstein says,

> There are two different conceptions about the nature of the universe – the world as a unity dependent on humanity, and the world as reality independent of the human factor.

This articulation seems to imply that these two conceptions can be placed side by side, compared and evaluated as to their validity. In other words, he seems to suggest that the two rival claims he is juxtaposing have their locus in a single framework. But there precisely lies the problem. Whether you think such a juxtaposition is possible or not depends on which position you take on the question itself. If you take the former position, there is no reason why that cannot be done. Whereas, if you are taking the latter position, you will notice that your position can never be articulated properly within those terms of reference.

So, Einstein's formulation foregrounds the matter of trans-paradigmaticity very well and forces us to take note of this particular aspect of the incommensurability that is involved in such debates. In order to perspicuously articulate his own position and clarify the difference between the two conceptions, Tagore would have to alter the very terms of reference. That is, he would have to insist that actually the two positions belong to two radically different ontological frameworks and consequently they cannot be addressed from the same vantage point. This fact (that is to say, their different views on the locus and status of the point of debate, as to whether a particular theoretical divergence can be described adequately within the frame of reference of a certain paradigm or not) indicates the gulf between their paradigms. There are broadly two different ways in which this matter can be presented. One is to say, as mentioned above, that the gap between the two paradigms comprises the vantage points from which the totality is viewed. The other is to point out that for one of the paradigms the difference is a matter of *fact, truth and proof*, whereas for the other paradigm it is a matter of *meaning, definition and conceptual structure*.

Now, to describe this mater in the context of the Einstein-Tagore dialogue, our task is to take both meta-positions into account and situate both conceptions in a common frame that has not already incorporated one of these positions. And in order to do that, we have to alter the terms of reference of both conceptions. This is not a matter of ambitiously positing a third, overarching paradigm as much as keep looking at each particular assertion from the global view of the other. In this particular case, a*t the level* at which Einstein is articulating his position, one could safely say, Tagore would not have found any reason to disagree. He is not advocating some near-solipsistic, anthropocentric ontology according to which the universe is the arbitrary projection of the human consciousness. Nor is his position incapable of accommodating the distinctions of objective and subjective, real and unreal, mental events and external world. Conversely, nor would Einstein have failed to appreciate Tagore's discomfort with the implications of a position according to which the universe is a meaningless assortment of particles in which the notion of human experience is totally unintelligible. But as a scientist, his *starting point* is the world as a given, and he cannot see how he can afford to give up that starting point.[1]

The nature of the conceptual divide becomes even clearer when Einstein protests that the view stated by Rabindranath amounts to 'a purely human conception of the universe.' As I pointed out above, in the sense in which Einstein takes Tagore to be saying it, the latter is definitely not saying that. What he is insisting on is the phenomenal character of reality. In that sense, he is saying that we do not know of any other conception of reality. That is what he means when he says, 'The world is a human world.' The word 'human' that Tagore is using to denote the phenomenal aspect does add to the confusion and he seems to be aware of it. He tries to mitigate it when he states that science comprises 'the impersonal human world of truths.' But without a proper gloss, this would sound as a vague, compromise stand in which human subjectivity is both asserted and denied as the ground of reality. And this kind of confusion occurs more than once in the dialogue.

The problem, as we noted earlier, is that for various reasons neither of them seems to have been able to see that the similarity of their vocabulary conceals a very fundamental conceptual divide. Rabindranath does give some evidence of an awareness of the nature of the miscommunication, but he is not able to – in any case, the context of the conversation would have probably made it jarring – offer a more perspicuous formulation that could expose the confusion. However, he refuses to yield his ground and only reiterates his position, almost in the same words. Hence our task to rearticulate their positions in an accessible and unbiased framework.

There are many ways in which one could look at the structure of this issue. One of them is in terms of the relation between ontology and epistemology – the ultimate in/dependence of ontology from/on epistemology, and in a related way, in terms of the necessary relation between realism and idealism.

5

One of the things philosophy has always forgotten and always reminded itself of is the priority of the epistemological. When we look at the world unreflectively, we find it to be already there. If we insist on the self-evident legitimacy of this consideration, in doing so we forfeit the right to engage in any kind of philosophical speculation. For the unreflective realism of the human organism is very different from the sophisticated realism that philosophers and scientists deal with. The latter does not spontaneously flow from the former, nor is continuous with it. In fact, it is doubtful if the former is even a good starting point to derive the latter. The realism of the scientist and the philosopher is ultimately the product of a reflective awareness that the existence of the world is grounded in our cognitive relation with it. This is the insight that has been, as I said, gleaned and forgotten, recalled and then ignored again and again, the latest reminder being Husserl's attempt to forcefully and comprehensively present this fact.[2]

Quite simply, one might say, it involves the question of the legitimacy of a certain question: after saying that something exists, is it legitimate to ask 'how do you know that it exists?' In other words, it is to do with the validity or otherwise of the claim that – after the decisive moment of reflective awareness – I have to *earn the right to affirm existence* and that the fact of my cognitive relation or my intentionality is what confers that right to me. Those who insist on the validity of this claim would argue that invoking the pre-reflective awareness of the world at this juncture is disingenuous.

The other, related way to look at this issue of knowledge and existence is, as I hinted above, to see that the trajectory of all human knowledge necessarily begins in subjective experience and ends in the transcendental positing of the world. We must begin in idealism and somehow end with realism. I must somehow reconcile the world being in my consciousness with my consciousness being a fact in/about the world. One could say that the entire history of philosophy can be seen in terms of the diverse attempts to perform this task in a way that does justice to these basic intuitions of our self and the world. It comes down to the question of Being: the status of the Being of things in consciousness can never be equated with the status of Being understood as existence; conversely, the status of the Being of consciousness can never be equated with the status of Being of things as existents.

In simpler terms, all the efforts to reconcile consciousness and the world, each as existing, have their point of departure in the acknowledgement of the fact that, to begin with, our reflection must start only in an awareness that our experience is all there is. In fact, this is so fundamental a gesture that we can just use the word 'reflection' to denote it. To reflect is, as such, to confront our awareness that in the final analysis, there might be nothing outside our experience, or rather, that there might be no exteriority to our experience. The ability of consciousness to posit an 'object,' its intrinsically intentional character is no guarantee that there indeed is an exteriority let alone an external reality. So, our ontological reflection must necessarily begin with the testimony of our experience on its own terms. There might be differences on the matter of how tentative is our acceptance of that testimony and how quickly or through how short a route we wish to reinstate the world more or less as we experience it in our pre-reflective intuition. But broadly, it cannot be denied that the available ground of being is coterminous with the circle of our experience. This is what, one would think, Wittgenstein meant when he said that the limits of my consciousness are the limits of my world. This is also the basic insight from which phenomenology starts.

6

Now having started from this point of departure, different philosophical theories move in different and quite divergent directions. But the crucial moment – *and this is the crux of the Tagore-Einstein discussion* – consists in the decision as to whether to leave the starting point behind or to retain it as a necessary element in our final understanding of the structure of reality. In other words, the question is whether, after reconstructing an independently existing reality from the contents of consciousness, it is legitimate to treat the bare fact of our experience as just another feature of that reality and our consciousness yet another (epi)phenomenon. Those, like Rabindranath, who regard the notion of an independent reality as a derivative that forgets its deriving source, would insist that the final description of that reality must continue to account for the fact that the condition of possibility of all phenomenal reality is our experience – another way of saying that to insist that the *phenomenal* nature of reality cannot be at any stage bracketed away. The circuit, so to speak, is not complete until our final description of reality coherently includes its initial emergence through our consciousness. The advocates of independent reality like Einstein point out that they are not guilty as charged since their description of reality includes human beings, their consciousness, their cognition and the conditions of validity of that cognition. But this, their interlocutor would say, is to miss the point.

As the phenomenologists have pointed out, you cannot posit consciousness as part of the external world. You may see its effects, the results of its agency, but you cannot see consciousness itself, at any rate in the sense in which you see the world. As Jean-Paul Sartre pointed out, there can be no coherent ontology in which Being is shared by consciousness as well as by things. If the latter represent Being, the former can only be seen as a negation of that Being, a gap in the plenitude of that Being. More pertinently, the insistence of the human-dependent reality proponent is that there is no legitimate gesture through which you can snap the link to the cognitive with which you began in the first place. For practical, survival purposes, our pre-reflective consciousness does repress that link. Having posited the external world, we do not want to be distracted by a constant reminder of its origin in our consciousness. When you are trying to locate the direction of some suspicious sound, it is not expeditious to become aware of how the 'sound' is actually happening in your ear. It is more useful to take the attitude that, 'the sound is coming from behind that bush, so I am able to hear it.' That is the reason why when Berkeley seems to say, 'the table exists because you are able to see it' or more accurately when he seems to say, 'the table exists only insofar as you are able to see it,' it sounds perverse. It would not sound that way if we were to allow reflection to remind us that 'the table exists because my consciousness has posited it there, and having been posited, I simply take it that the table is there and I can see it.' In other words, it would be easier to understand this point if we were to remember that *we owe our ontology to epistemology* or to look at the matter from another perspective, that the edifice of our realism and its comforting stability rest on the foundation of idealism which is not visible.

One way of making that foundation visible is to look closely at the primary metaphysical binary of appearance/reality, the idea that what is perceived by our consciousness is appearance and that there is a reality that is distinct from it and independent of it. Now, we often fail to notice that the notion of appearance is very slippery. The first problem is, how exactly do we know that a particular reality (assuming that we have reliable access to it) corresponds to a particular appearance? What is the status of that link, what is the vantage point which allows you to see it? That is, where exactly are you standing, and what faculty are you using when you see both the appearance and its corresponding reality, and can see that they correspond to each other? If we have no such vantage point, what is the status of the notion of appearance? If it cannot give an account of itself as a term of a binary, is it not self-negating? Given this difficulty, what justifies using the notion of appearance to posit 'reality'? Does it not make more sense to throw away the entire package comprising the dichotomy of appearance and reality?[3]

7

To take the matter to its next level, the issue, if one may put it simplistically, borrowing Husserlian terms, is about the *status* of intentionality. This relation is the bridge between the already existing consciousness (since that is our point of departure – not withstanding all the critiques of Descartes) and the yet to be posited (and eventually transcendentalized) world. The question is whether we can withdraw that bridge back into consciousness and make it another element of consciousness or push it out into the world and make it part of the furniture of the world. The adherent of the human-dependent reality position would argue that both actions are illegitimate: the former attempt would end up in psychologism and the latter into a disingenuous form of realism. The bridge must retain its status as the bridge between consciousness and the world. If it results in a paradoxical, Escher-esque picture in which the world is in the mind and the mind is in the world, so be it. At least it is an attempt at a complete picture, at any rate it is an honest picture.

This is the subtext of Tagore's insistence that

> the infinite personality of man comprehends the universe. There cannot be anything that cannot be subsumed by the human personality, and this proves that the truth of the universe is human truth.

To elaborate, he is asserting that our conception of reality cannot be disengaged from the fact that it is *our phenomenal conception*. Second, he is insisting – and this becomes more explicit later on – that not just the analytic mind of the human being but her entire personality is implicated in this. (We shall return to this point later). Third, he is arguing that we must look at the 'human' pole of this fundamental relation not in terms of the contingently human but the essentially human. He articulates this more clearly when he states that

> science is concerned with that which is not confined to individuals; it is the impersonal human world of truths.

In other words, for the stability and coherence of our vision of the world from which we draw nomological inferences and posit laws of the universe, we are not invoking some particular individual's cognition but human cognition per se. The logical conclusion of this position is what Rabindranath is asserting when he refers to *superpersonal man* and *universal being*. If any sense of reality that we can conceive is necessarily phenomenal, then in order to coherently (that is to say, in a way that is really intelligible to us) posit the world as it is in itself or as an absolute reality, we must retain that phenomenal character. We cannot pretend to suddenly have access to, or even be able to make sense of, a non-phenomenal reality. This is what Kant so emphatically pointed out. If this is what we need to do, in that case we must retain the phenomenal character for what we all call objective reality but conceive it in terms of an all-pervasive, cosmic intentionality instead of the finite, localized intentionality of human consciousness. This was the insight Berkeley tried to capture in his claim that since *esse est percipi*, only the cognition of God can be the locus of the reality of the world. It must be understood that the 'God' Berkeley is invoking is not a Creator God who answered to petitionary prayers. His insistence is that coherence demands that we remain within the phenomenal frame and this in turn requires that we posit a cosmic consciousness in terms of whose intentionality we can speak of reality. That reality is doubtless 'independent of our human cognition,' but it is not outside cognition. In other words, cognizability is the space in which alone reality can be conceived. Tagore puts this explicitly when he says,

> [an object, like a table] remains outside the individual mind, but not the universal mind. The table is that which is perceptible by some kind of consciousness we possess.

That is, the table we see is doubtless a version limited and distorted by the constraints of our human consciousness. But the table as it is in itself cannot escape the condition of phenomenality. This point, which Berkeley made implicitly, Kant made in a more unequivocal fashion. But this leads to another extremely important and troubling question.

How exactly is one to make the leap from the particularity of the finite human consciousness and its limited phenomenality to the phenomenality of the cosmic? True to his native tradition and his own personal convictions, Tagore states that

> we realize the supreme man, who has no individual limitations, through our limitations.

We must, he implies, retain the *structure* of that phenomenality. We have to conceive of the source of that phenomenality in analogy with our own consciousness. But one might ask, how justified would we be in holding that analogy? Do we have any grounds for it? How do we know where the essential structure of our finite consciousness ends and its merely contingent character begins? The answer is we don't know. But if it has to be consciousness of some sort, it must share something with our own finite consciousness and we can only make a plausible conjecture that after we eliminate the merely psychological, what remains must be the essential structure of consciousness. This is what Tagore is hinting at when he refers to 'realizing the superpersonal through the limitations of the personal.' This is in fact the direction Husserl took. One could always argue about the merely speculative character of this approach. But the important point is that no one has been able to suggest an alternative conception. From Plato to the present day, what have been offered as alternative conceptions are simply different variants of the same argumentative structure. Some people speak of the cosmic in terms of an 'impersonal intelligence,' but as the *Advaitin* would be quick to point out, it is indeed legitimate to do so but we must realize that at that level, ontology contracts to a point of singularity. There would be no room for any kind of statements or descriptions. There would be neither ontology nor epistemology.

8

Einstein is of course quite justified in not being interested in this aspect of the matter inasmuch as he is not required, as a scientist, to take this into account when he endeavours to give a coherent description of the relations between various phenomena. His task does not require settling the question as to the ground of the totality of phenomena that we call the cosmos. The problem is that the average scientist (though it hardly needs to be added that Einstein was a remarkable exception in this regard) does not realize that he is operating with an incomplete and ultimately incoherent frame, and that this is permissible only because his concern is with the *inter-relations within that frame*. He rashly assumes that the cogency of his account of these inter-relations is sufficient proof that the frame is complete and coherent.

That is, the scientist does not realize that in the limited, implicitly phenomenal sense, in which he claims objectivity and independence to reality, it is not contested by the philosopher. He does not see that the philosopher is only asking him not to wish away the frame in which his limited notion of posited reality is entrenched. More pertinently, the philosopher's insistence is that the notion of 'existence' ultimately comes from a certain manoeuvre of our phenomenal experience. This is of course what Derrida, following Heidegger, called the Metaphysics of presence. But the fact remains that it is the only kind of metaphysics we can conceive. Either we acknowledge our immersion in this metaphysics and remain vigilant as to its implications or we take the positivist route and treat existence as a convenient fiction, or at the most as a necessary assumption. In that case we forfeit the right to speak of reality as such.

The immediate consequence that follows from this is that reality is a matter of correct perception; it is that which answers to truth (understanding truth here as the character of our relation to reality). This being so, the notion of reality must retain its relation to truth, only transcending from human truth to a cosmic truth. And to speak of truth at that level is to posit a commensurate consciousness. Tagore makes this point clearly enough when he says,

> what we call truth lies in the rational harmony between the subjective and objective aspects of reality, both of which belong to the superpersonal man.

The only problem here is that the terminology Rabindranath uses tends to give the impression of an anthropocentric view of the universe. All that is needed, in order to avoid misunderstanding is to keep in mind that these notions such as 'personal,' 'man' etc. are just a way of referring to the intentional character of our sense of reality.

9

Now, moving on to the next crucial point, let us look at what Einstein says. He says, (speaking presumably on behalf of scientists)

> We attribute to truth a superhuman objectivity.[4]

There are two terms here which need to be examined closely. First, is the notion of truth, which we already had occasion to talk about. What exactly does Einstein mean by the objectivity of truth? What kind of thing is truth supposed to be in this context? It is, I think, obvious that he is invoking one of the forms of the correspondence theory and saying that we are capable of perceptions which are true by virtue of their correspondence to a reality which is not dependent on us for its existence.[5] So, the picture he is trying to give is of a world that is already there, which we perceive and when we perceive it correctly that perception embodies truth.

Rabindranath would make two points in response. First, Einstein's conception of truth itself necessarily requires the notion of human consciousness for its coherence (unless he is using 'truth' as a loose synonym for 'reality'). Second, Einstein's concept of truth already presupposes a reality in whose existence (and accessibility) our consciousness plays no role. He would object that it is not legitimate to invoke a notion of truth that already presupposes a certain conception of reality in support of that conception. The more proper way to construe the notion of truth would be to ground it in a movement *from truth to reality* rather than from *reality to truth*. In taking the latter option, Einstein is already assuming that he knows a certain thing to be reality and all that remains for him is to check whether a certain perception of it (or an assertion representing it) is true or not.[6] The question is how did he know that the thing in question is the reality? How did he know that he was not a victim of error?[7] Evidently, Tagore would say, Einstein used his perception and relied on the truth of that perception in order to identify that thing presented to his consciousness as reality. Reality may confirm the truth of a perception, *but before that*, that reality was identified and confirmed as reality by that very perception. 'Reality' owes its status to the 'truth' of perception.

The second term in Einstein's statement is 'objectivity.' Tagore's response to this part of Einstein's statement would be that, contrary to the latter's claim, truth is not a matter of superhuman objectivity but of superhuman *subjectivity*. He would have added that it is futile to give any other kind of sense to the notion of objectivity. From the point of departure we have referred to, objectivity can only be a function of intersubjectivity. Einstein goes on to say,

> It is indispensable for us – this reality which is independent of our existence and our experience and our mind – though we cannot say what it means.

What Einstein is accepting here is that he is aware of the ambiguity – in fact the paradox – of the notion of a *mind-dependent sense of a mind-independent objective reality*. As he frankly admits, we do not know the exact status of such a claim. At the most, all we have is a dogma that our consciousness has such a capacity for transcendence. That it is a dogma and no more is clear in Kant's account of what he calls 'the unity of apperception.'

But before moving on to the next aspect of the matter, let us revisit the scientist's claim that his task is to describe the relation of phenomena or more accurately just the continuous structure of phenomena, and for this he requires the assumption of an independently existing reality. There is an important factor here on which in fact the entire business of metaphysics rests. It relates to the question of what *description* means. What exactly is it to describe the structure of phenomena?[8] If we look at the notion of description carefully we realize that it is a deeply phenomenal notion. Description is the absence-version of perception. To put it simply, when you can perceive, description is redundant. Description is a substitute for perception for those who are not present in order to perceive. In that sense, all description is, directly or indirectly, a form of recognition. Therefore, it follows that since perception is phenomenal, description must be phenomenal too. A description must be couched in terms of perception and must be intelligible to the absent perceiver. At this point, the scientist – particularly the physicist – is likely to point out that he does not deal in that kind of description. His description consists of mathematical structures. His language (of description) is not the language of sense perception. It is the language of mathematics, which is not a matter of perception but of logic and Reason. In doing this, whether she is aware or not, the scientist is opting for the quintessentially metaphysical dichotomy of *the sensible versus the intelligible*, and casting her lot with the latter. This is the rationalist stand since Plato, valorizing Reason at the cost of the senses, with its implicit assumption that the world of forms transcends phenomenality.[9] However, the precise nature of the relation of mathematical descriptions to their supposed objects is far from clear. As pointed out above in a related context, unless an extreme operationalist and instrumentalist view is embraced, thereby preempting that question itself, it is difficult to explain how exactly the scientist establishes that a certain mathematical description correctly fits a particular phenomenon. To put it in simpler terms, whatever the autonomy of a world of pure forms – and by implication mathematical structures – our concern is with the application of those forms to phenomena.

To return to the larger point, our efforts to establish the relations between structures of phenomena based on their past and projected behaviour, are entirely within the phenomenal frame. To deny this is to fall into a mistake analogous to the error of imagining that primary qualities, *while inhabiting the same world as the secondary qualities*, are objective while the latter are subjective.[10] As several philosophers have pointed out, the space, time, entities and relations posited by the physicist are in a complex relation to the world of our experience. It is not at all clear what we should take to be the status of a statement like, 'impenetrability *is the same as* repulsion of atoms' or 'impenetrability *is the effect of* repulsion of atoms,' and how exactly we should understand the part placed in italics. The question is whether these statements are coherent in the strict sense of whether all their constituents and their governing syntax are from a continuous conceptual space.

More pertinently for our discussion, the point is that a metaphysics that operates with a phenomenal/transcendental binary is as guilty of inconsistency as its other variants. In other words, in speaking of structures or forms we may imagine that we are replacing the perceivable with the conceivable. But as Merleau-Ponty pointed out, to reduce the perceivable to the status of an object of the conceivable is to miss the former completely and only clutch at its shadow: a colour is not the same as the idea of that colour or the quantitative expression of its wave-length and frequency that denote its structure. In short, experience is irreducible.

10

Let me now briefly talk about the issue I promised to return to. This is the question of the *personality* of the human as well as the superhuman being. The point Tagore is anxious to make in the dialogue is that we must liberate ourselves from the idea that even cognitive consciousness is some sort of a pure, non-affective, analytic mind. This is the fallacy in the Cogito of Descartes, which Existentialist philosophers and others, such as Feminist thinkers have pointed out. First of all, such a picture of human consciousness is simply incorrect. Each advance in the field of cognitive science makes us more acutely aware of this fact. This fallacy moreover has serious consequences. The assumption that the perceiving consciousness is a neutral, inert faculty leads to a notion of truth, knowledge and the reality of the world that is suffocatingly restrictive.[11] If we were unemotional beings who did not know the meaning of love, suffering or compassion, who did not respond to beauty and were incapable of dreaming, the world we posited through our consciousness would be an unrecognizably different world, a dreary, desiccated world perhaps not worth living in. It is not just our senses and our analytic mind that participate in the construction of the world through our consciousness. *Our entire personality, the totality of our being participates in the construction of the world.* In that sense, our world is permeated with our personality and reflects the texture of our interiority. This fact has enormous bearing on our lives much more saliently than we imagine. Let us touch on this aspect of the matter by asking a question about the larger significance of this dialogue between Einstein and Tagore.

What is the importance of this dialogue, for that matter, this entire issue? Does it really matter what view we take – whether reality is inalienably human or is independent of our minds and our personalities? The scientist goes on with her work of explaining the structure of phenomena, the mystic is absorbed in his attempt to unite with the Absolute and the rest of us trudge along somehow. So what is at stake here? The answer is, everything – everything that is valuable.

Once we suppress the phenomenal frame of reality and posit it as independently existing and endow it with only those attributes that are accessible to the neutral, non-personal, cognitive mode of consciousness, what we have is a world full of inert facts[12] with a total absence of anything even remotely resembling values. It is a world in which human life is an accident and the human mind is the epiphenomenal mist surrounding it. It is a world in which there can be no distinction between right and wrong or good and bad. It is a world in which there is no reason to prefer good over evil, love over hate, beauty over drabness, compassion over cruelty. All you have, at best, is the ambivalent solidarity of millions of organisms all struggling desperately to continue their DNA in the face of the inexorable force of the second law of thermodynamics. In such a world, to pretend that our norms, values and ideals have any basis, any meaning at all is duplicitous. Religions, even if they were grand lies, posited a soul which was a vehicle of meaning, and a god who amidst all the cosmic meaninglessness was the source of value. But not anymore. This is related to what Max Weber called the disenchantment of the world. The disenchantment is not just about the vanishing of the sense of mysteriousness of the universe, where everything is exposed to the unrelenting glare of scientific, rational explanation. Unfortunately, there is more to it than that.

The really serious dimension of disenchantment consists in the fact that there is no longer any ground for values. The result has been that most values have become quaint habits anachronistically retained from the past. The only value that has remained relevant somewhat is 'decency,' but we do not know why anyone should bother to be decent. Generally, science has been held responsible for this state of affairs but the truth is that scientists qua scientists only took on board what they needed and what they could deal with. If anyone, dazzled by the spectacular achievements of science, accepts its underlying, deeply problematic metaphysic as a matter of blind faith, they have themselves to blame. It is not science per se but scientism that is the culprit. As a rule, responsible scientists have not been eager to promote a particular worldview, even one based on science. If anyone, it is the scientifically minded philosophers who have done that. And it is these philosophers who must answer all those whom they led into accepting this scientific, rational, mind-independent picture of the world.

It is not that the scientists and philosophers are not themselves victims of this predicament. They are. But the scientist deals with the situation with a special sort of schizophrenia which he has honed into a fine art over the last few centuries. But the philosopher who brought about all this in his anxiety to put philosophy in the service of science and use it as the foundation stone for the kind of 'self-evident,' 'indubitable' foundations that science was supposed to be resting on, it is he who finds himself with no recourse. No wonder he has been looking distinctly embarrassed. And he ought to. For it was he who, in order to avoid the fuzziness of metaphysics, borrowed the implicit positivism of a dominant ideology of modern science (and along with it, its more confused metaphysics), and sold us all this worldview, and worse still, committed the cardinal stupidity of believing it himself.

Now by way of making amends, the modernist philosopher is trying to re-enchant the world – without of course using such dubious things as magic, mysticism or religion. As part of this good gesture, he has been trying to rehabilitate values in the world. The tools he has been deploying range from the glib argument that the fact of human agency proves that the world is imbued with values to the somewhat convoluted claim that values are a sort of secondary qualities.

11

As for the claim that agency constitutes proof that the world has a dimension of value, I think it is based on faulty reasoning. This is not the place to attempt a proper refutation of the claim, but we can look at the basic contours of the argument behind it and see why it is not as sound as its proponents might think. The fact is that contrary to what this view assumes, agency can only prove the *need* for value. It cannot prove its existence. What characterizes the world is the fact that there is a plurality of things that are different and unequal. Given the fact of agency, this *enables as well as necessitates* choice. Say there are two things that are not the same or equal in certain respects and you have to choose between them. It is of course assumed that since you are endowed with agency, you can choose. However, the fact that you have choice does not imply that the situation of choice provides you with value. The difference and inequality of things does not come with a value-code inscribed in them. One of the two things before me does not present itself as having greater value than the other. Their difference is just difference and nothing more. I have to create a system of value and assess the two things on the criteria provided by that system and pronounce one of them as more valuable than the other in one or more respects. I have to use value to make the choice in a particular way. But value as a discriminating principle, that is to say, value as that which gives a normative structure to the difference, does not present itself to me. *I* have to contribute it to the situation. On the basis of my beliefs and desires (which do not emanate from the objects or their differences) I have to create a system to order the differences in alignment with my desires. This is the point the existentialist philosophers were making – whatever the worth of the solutions they were offering or were refusing to offer. At the most, therefore, what we could say is that, what agency and the plurality of things provide are the conditions of possibility and conditions of necessity of value, which is very different from the existence of value in the world.

However, the relevant point in this context is that, whether this claim can coherently be made at all or not depends on whether it is made within the human-independent reality framework or the consciousness-dependent reality framework. If human consciousness is an entity – even if a rather peculiar one – among other kinds of entities in the world that is posited independent of it, this view either makes little sense or if you try to strenuously make it coherent, it just becomes false in a quite manifest way. The only way this view can be coherent and meaningful is if it is an extension of an unapologetically phenomenological framework with all its idealist implications. In a universe that is necessarily phenomenal, with a superhuman consciousness as the locus of phenomenality, agency would imply value as an integral dimension of the world. But that would be a frankly religious world, and not the kind of world Einstein is talking about in this dialogue.[13] The philosophers who have been attempting to advocate the view we are discussing, do not seem to be aware of this fact. If the philosophers in question realize that in order to be coherent, their argument will have to be situated in a phenomenal conception of reality, they will have to pose the problem itself in a quite different way, since they are no longer trying to reinstate values in a disenchanted world, a world that has revealed itself as existing independent of and unrelated to human Being.

The other attempt consists of the claim that values are part of the world and are not mere human constructs, but we must think of them along the lines of secondary qualities. John McDowell has in particular tried to present and defend this view in a fairly rigorous and persuasive way. Unlike the agency-proves-values position, this view accepts the phenomenological frame in a more conscious way. But there lies the problem. As the history of phenomenology shows, it is futile to try to weld phenomenology with some other framework or to mitigate it with some concessions in order to rid it of its idealistic implications. In any case, McDowell and other philosophers who favour this position are from the analytical tradition and their comfort zone, for that matter their default ontology, is the mind-independent realism compatible with the implicit worldview of modern science. Why this point is important here is that the status of secondary qualities is different in a phenomenological frame from its status in the ontology of scientific realism.[14] In the former, the difference between the two would be a matter of degree. In the realist framework, on the other hand, secondary qualities in *conjunction with primary qualities* are – as suggested earlier – a completely incoherent proposition.[15]

As with so many other issues in metaphysics, this has to do with the relation between qualities and relations. As the idealists never tired of pointing out – even at the risk of being ridiculed – the notion of qualities per se is problematic. Strictly speaking, qualities as such can be posited only in a reality that is supposed to be independent of consciousness – a fact which in a *reductio* sort of way proves the untenability of that metaphysics. But even in that schema, qualities can be understood only in terms of their mutual relations, that is to say, in terms of their mutual effects. But qualities and relations together cannot be described coherently except in relation to consciousness.[16]

Given this background, the suggestion that values are (or are akin to) secondary qualities is not a very useful strategy if the aim is to argue that values are objective. After all, as said earlier more than once, Berkeley's approach – which was only taken to its logical conclusion by Hume – was that the notion of secondary qualities betrays the inconsistency of the notion of primary qualities and that therefore the metaphysical frame that Locke operated with needs to be revised in such a way that all qualities are essentially secondary qualities. If this is the implied stand of the philosophers who try to argue that values are secondary qualities, then they must realize that this is incompatible with the kind of scientific realism which they broadly accept. In the latter, the world is basically constituted by primary qualities and is mind-independent. To reiterate what we have said already, there is as yet no coherent metaphysical frame in which both primary and secondary qualities can co-exist, and it is doubtful whether there *can* be one.

That some confusion exists in the minds of the values-as-secondary qualities theorists on this count is evident from the way McDowell for instance tries to make use of the phenomenological frame in a partial sort of way without committing himself to its full implications (which as I pointed above, is just not possible). If we may telescopically compress the matter, the inescapable fact is that secondary qualities imply a phenomenal view of reality. If we are willing to take that step, then not only values but facts as well have the ontological character of something like secondary qualities, or to put it more accurately, they will reveal their phenomenal character in a more manifest fashion.[17]

Anyway, the larger point is that the scientific worldview as such cannot provide a conceptual space for values, ideals or human meanings, except in a totally reductionistic way in which all these things are reduced to instincts, drives, hormones and chemical interactions. The attempts of philosophers to forcibly create a conceptual space in that schema has not succeeded and is, for the reasons briefly cited above, unlikely to succeed. The implications of reductionism cannot be wished away with some philosophical legerdemain. You cannot suggest that values and our concern for them are nothing but survival strategies or the doings of some specific gene or some such thing, and in the same breath ask people to take values seriously and live by them. There is a tendency to imagine that these are all settled issues and that there is no need to rake them up all over again. The truth is that these issues have not been resolved in any satisfactory way. What prevails is a somewhat shallow form of pragmatism in which you pretend that the incoherence of our worldviews does not matter since we have scientifically and technologically moved so far ahead that this shortcoming cannot affect us in any way. But we ignore the inconsistencies in our worldview only at our own peril.

Surely it must have been a recognition of the gravity of this matter that prompted these two great thinkers to spend a part of the all too brief time they had at their disposal on this topic. We should be grateful to them for the insights their conversation offers and its thought-provoking potential, thanks to which we find ourselves engaging in this discussion.

Notes

1 Husserl's ambition was precisely to provide an alternative starting point that is more coherent in the larger perspective. But, Einstein was not aware of Husserl's phenomenological initiative, and it is difficult to speculate as to how he would have reacted to it.

2 When we say 'the latest' we do not mean to deny the great advancements and refinements beginning from Merleau-Ponty, Heidegger and many Post-Heideggerian thinkers. But for all of them, one could fairly say, Husserl's work remains the point of departure.

3 To see the point of these objections is to move from metaphysics to ontology proper. But to pursue the structure of that movement in detail here would be a digression.

4 What he means of course is 'human-independent' objectivity, but we need not be distracted by this minor matter.

5 His emphasis here is not on the point that the notion of a subjective truth is incoherent. That he would in any case have emphatically maintained. We are past the vulgar misunderstanding that Einstein was relativist.

6 The aim of Heidegger's discussion of the concept of truth in terms of *aletheia* and *unconcealedness* is to point this out. His main point on this question is that when we speak of truth, and try to establish it in terms of correspondence, we assume that facts are already out there and all we need to do is to verify our propositions against them. His suggests that unless, prior to our verification, we ask how the facts came to be there in the first place, our understanding of truth will remain vitiated.

7 This, by the way, explains why classical Indian philosophers focused so much on the question of error in their epistemological enquiries.

8 Let us sidestep the claim that the job of the scientist is to *explain* phenomena. We need not go into the details here but a little reflection will show that explanation ultimately must resolve itself into related descriptions. Therefore, the scientist is well advised to stick to his claim of describing the structure of phenomena.

9 But ever since the ambiguities that underlie this primary structure of metaphysics have been rigorously exposed by Heidegger and later by Derrida, rationalism finds itself much less invulnerable than it used to.

10 It is a great pity that one of the most cogent and compelling expositions of this matter by A. N. Whitehead has fallen by the wayside. He pointed out the fundamental incoherence of the general pattern of this position through reference to what he called 'the fallacy of simple location' and 'the fallacy of misplaced concreteness', in a way that is most instructive.

11 This is the direct consequence of the recourse to mathematical description as the ultimately reliable account of realty, which we discussed a while ago.

12 Even this is not strictly speaking, accurate, since 'facts' too are a contribution of our thinking. When Wittgenstein said that the world is a collection of facts, his statement makes sense only in a phenomenal frame. But this is a large issue and we cannot go into it here.

13 Elsewhere Einstein emphasized the importance, indeed the indispensability of such a view. As a matter of fact, one could describe that to be his preferred view. But here we are only talking about the view he expressed in this dialogue.

14 I find myself forced to repeat some of the points made earlier with regard to secondary qualities. I apologize to the reader for the repetitiveness.

15 After all, this is the insight that constitutes the point of departure for Berkeley.

16 It is in this sense that the difference between primary and secondary qualities is, in the final analysis, a matter of degree. We know all qualities only in terms of the effect they produce in consciousness. To say this is not to conflate them with sensations. Sensations are rightly regarded as dead ends. But qualities as existing by themselves – regardless of their relation to any substance – cannot be conceived.

17 In fact, a number of philosophers in the analytic tradition have been alert to this implication for quite some time.

References

Derrida, Jacques. 1981. *Dissemination*, trans. Barbara Johnson, Chicago: University of Chicago Press.

———. 1978. *Writing and Difference*, trans. Alan Bass, Chicago: University of Chicago Press.

Heidegger, Martin. 1962. *Being and Time*, trans. John Macquarrie and Edward Robinson, New York: Harper & Row.

Husserl, Edmund. 1970. *The Crisis of European Sciences and Transcendental Philosophy*, [1936/54], trans. D. Carr, Evanston, IL: Northwestern University Press.

———. 1960. *Cartesian Meditations*, 1931, trans. D. Cairns, Dordrecht: Kluwer.

Merleau-Ponty, Maurice. 2012. *Phenomenology of Perception*, trans. Donald A. Landes, New York: Routledge.

Whitehead, Alfred North. 1933. *Adventures of Ideas*, New York: Macmillan Company.

———. 1929. *Process and Reality: An Essay in Cosmology*, New York: Macmillan Company.

Wittgenstein, Ludwig. 1961. *Tractatus Logico-Philosophicus*, trans. D. F. Pears and B. F. McGuinness, New York: Humanities Press.

10

REALITY AND INTELLIGIBILITY

Bijoy Boruah

The real and the truth

Granted that *Reality* is an ontological category and *Truth* is (in some sense) a matter of correspondence with Reality, it is a moot question whether the required *objectivity* of Truth implies, or demands, its recognition in an *epistemology-transcendent* mode. If assertions of truth about reality are considered to be dependent on 'ways of knowing' what there is or what the world is like, then the objectivity of Truth (i.e. how the world is *anyway*, as Bernard Williams would put it)[1] may not be attained inasmuch as the particular epistemological means or modes (e.g. human or otherwise) underpinning the assertions may lack objective adequacy. On the other hand, were it not for the fact that there is some epistemological route or other that has the prospect of actually meeting with reality, there would be no concept of truth.

What essentially underlies the problematization of the relation between Reality and Truth *vis-à-vis* the human conception of the world in the above way is the seemingly unbridgeable gulf between *metaphysics* and *epistemology*. It seems that metaphysics, or the theory of reality, commands independence from epistemology, because what is (metaphysically) real is so irrespective of whether we know it or not, or how we (claim to) know it. But it is equally pressing upon us, on the other hand, to believe that a metaphysical *conception* of reality is, after all, an account of reality as *intelligible*. And *intelligibility* necessarily invokes the notion of *intelligence* (or consciousness/mind). The problem about 'Reality and Truth' is therefore one of formulating the idea of *metaphysical intelligibility*. The crux of this formulation is the requirement that the intelligence in question has the right measure of objective adequacy. It would then be the conception of an *ideal* intelligence that is intrinsically capable of transcending all limitations or deficiencies immanent to any perspectival representations of reality.

True because real: a realist prejudice?

There is a clear sense of *conceptual* priority of the real over the true. We would talk about *this* or *that* being true only because *what* is asserted to be true is identified with, and recognized as, real. Indeed, this is what the proverbial 'true to the fact' is taken to mean. And an assertion is true in the plain sense of being *made* true by the fact that it asserts, in case there is a fact of the matter about which the assertion is made. True assertions 'track' the truth just when they are made true by the fact in question. It is in this sense that truth is equated with the fact or reality. But this equation is not supposed to overlook the point of difference between the two concepts: truth and reality.

The fundamental difference is that the question of 'true or false' would not arise if it were not for the sake of determining what the real is. It would be quite odd to say that something is real because truth, or finding it true, has made it real. What this oddity implies is that the predicate 'being true' has no application without reference to what is real. And *reference* to reality, which is a paradigmatic mode of coming to terms with truth about the world, is obviously a genuine act, and an actual semantic concept, only because reality provides the context or basis for its genuineness or actuality. Idiomatically put: 'No reality, no truth, or no *context/basis* for truth.' But the converse of this idiom does not hold; for truth is not the context for, or basis of, reality.

Truth has to correspond (in some sense) to reality in order to be *counted* as truth; but reality does not have to correspond to truth, or to anything for that matter, in order to earn the status of reality. Indeed, something's being real is not a *status* the thing acquires or has, because the thing does not become real by somehow gaining any status. That Mount Everest is a real mountain does not mean that it had to earn its reality by virtue of the process of evolving into a mountain, even though its geophysical evolution is a truth about its reality. But we, conscious beings with cognitive endowment, discover this truth about Mount Everest. And our cognitive discovery – which might be characterized as our cognitive negotiation with reality – is registered in a celebrated assertion which is true to the fact of this geophysically evolved entity or phenomenon.

Truth is, thus, subservient to reality. If so, then reality is independent of the concept of truth. If something is real, it is so independently of any cognitive discovery or determination of the truth about this reality. That a real horse is real is self-evident, and also self-evidently true. A winged horse (unicorn), on the other hand, is unreal, but not because it fails to gain the status of reality. It cannot have or gain the status of reality, any more than a real horse can possibly lose the status of being real.[2] If, by any chance, a horse metamorphoses into some other kind of organism, then it would not lose the *status* of horseness. It would simply *cease* to be a horse by turning into a quite different organism.

The logic of this argument about the priority of the real over the true is that reality is conceptually self-sufficient but truth is not. What we take to be true is true *relative to* what is real, whereas what is recognized or acknowledged as real is real relative to nothing, and certainly not relative to our recognition or acknowledgement of its reality. It is *true* that there are horses in the world *because* there really are horses in the world. But it does not follow from this that there *really* are horses in the world because it is *true* that there are horses in the world. Even though it is true that there is X, that there is X (or X is real) is not *consequent* upon that truth. If X is real, then there is a discourse, a discursive statement, about this reality in the form of the following acknowledgement: 'It is true that there is X.' And to say that X is real *because* it is true that there is X is to put the cart before the horse. Truth needs to be accounted for by reference to the intellect of the cognitive subject; reality doesn't. Such is the asymmetry of the relation between reality and truth. Is this alleged asymmetry incontrovertible?

Reality and truth: a symmetrical relation?

If what is real is the only thing that can be said to be true, then it also implies that only what can be said to be true is real. If indeed X is real (or there is X), then *that* X is real (or there is X) is a truth. The propositional *that*-clause represents what is true, i.e. what indeed is the case. It also means that the proposition represents a *truth*, and the truth is the reality in question. More precisely, the truth *is* the reality. Thus analysed, the relation between the real and the true does not seem to be asymmetrical. It rather looks as though there is an *identity* between reality and truth. We should, in that case, be able to make an intersubstitutable use of these two expressions. However, we must then be willing to countenance truth as a *metaphysical* category or some kind of an 'ontic' notion.

In point of fact, we *do* use the notion of truth in the ontic sense whenever we talk about our quest for truth. Our journey in search of reality is our quest for truth. To know the reality of matter, mind, self, free will, justice, equality, happiness or anything is somehow to attain the truth about these things. It is to know the truth about these concepts, or about the content of these concepts. In this sense, the intent of the locution 'there are truths' is both legitimate and quasi-ontological. Truth would then no longer be just a semantic predicate attached to a sentence; it would rather be a quasi-ontological category distinguished from the epistemological process of coming to cognitive terms with it. Truth would therefore inherit the property of objectivity intrinsic to the concept of reality. Evidently, the objectivity of truth would then be metaphysical rather than epistemological objectivity.

Even if truth is objective in the metaphysical sense, there must be a *principled* connection between truth and knowledge. When it is held that truth is a necessary condition of knowledge, it is implied that whatever is the truth in question is *known* to be the case (or true). The so-called 'unknown truth' is thus a misleading expression because of its imprecision. What it precisely means is 'unknown reality' or the reality that is yet to be comprehended. That nothing moves faster than the speed of light is a truth (if it *is* true) precisely because it is known to be so in accordance with the most appropriate and hitherto unchallenged theory of physical reality. This does not mean that the concept of truth can be reduced to the concept of knowledge. It also does not mean that the concept of knowledge has logical primacy over that of truth. Rather, the point is that knowledge and truth are correlative concepts. Knowledge entails truth, but truth 'comes to light' only because that light is *knowledge*.

Characterizing truth as the *content* of knowledge generates a very important question. If the *raison d'être* of truth is solely its being the content of knowledge, then it looks as though truth is *dependent* on knowledge. But the alleged objectivity and autonomy of truth would be lost if truth is claimed to be knowledge-dependent, i.e. dependent on the knowing subject. Since the concept of truth cannot afford to lose objectivity and autonomy, and yet still truth cannot be truth unless there is the light of knowledge, there seems to be a dilemma that makes the problem serious.

Revelation of truth

Granted that truth is the central content of knowledge, the ascription of truth-content to knowledge also gives the impression of placing truth within the spectrum of epistemology. In other words, there is no occasion for (mention of) truth until and unless reality is revealed by the light of knowledge. Since knowledge is true cognitive representation of reality, truth is attained because the representation in some sense *captures* reality. Only when reality comes under 'epistemic capture' does knowledge occur to the cognizing subject. Only then does truth figure as something that the knower comes to attain. Hence the concept of truth partially owes its meaning to its essential link with epistemology. Truth is, we might say, *epistemically revealed* truth.

Despite its essential link with epistemology, truth is not an epistemic concept. Rather, it is an epistemically *qualified* concept, but this qualification is no reason for subsuming it under the epistemological category. What the qualification means is that the discourse of epistemology is a conceptual prerequisite to the 'emergence' of truth. There is no possibility of truth without the possibility of knowledge of reality. But it does not follow from this that there is no possibility of reality without the possibility of knowledge of reality. Epistemology does not, and cannot, set the conditions for the possibility of reality in the way logic does. (E.g., logic forecloses the possibility of a unified spatial region which is both red and green all over at the same time.) This is where 'truth' and 'reality' come apart in a significant way. Reality awaits epistemic capture and it may never come under such capture. But there is nothing incoherent in the idea of a reality beyond any possible epistemic capture, even though such an idea is quite likely to be vague or indeterminate.

A consequence of the significant difference between truth and reality is that there can be unknown reality but no unknown truth. *No unknown truth!* 'To be *true* is to be *known* to be true' sounds like an unacceptable idealistic dictum. It might sound strange to us to be told that all truths are *known* truths, which in turn would imply that the notion of 'unknown truth' is incoherent. What seems particularly unconvincing is the consequence of accepting this dictum, which is the rejection of any statements of truth in the future tense. If, by definition, truth is what is known to be true, projecting hitherto unknown, yet-to-be-accessed truths of the future would be misusing the concept *truth* by placing it outside the epistemic spectrum. But who would want to deny that there are more truths in heaven and earth than what we mortals already have in the basket of knowledge? Who but the utterly irrational would not have the intellectual humility to acquiesce in the fact that the set of truths incorporated in our knowledge is miniscule compared to the infinitely larger set of unrevealed truths still concealed in the darkness of ignorance?

Truth as the 'Measure' of reality

Astonishing though it sounds to us to be told that truth is by definition *known* truth, the epistemic qualification of truth is not at all wrong.

Of course what we know is reality, which is revealed by the light of knowledge. But it is precisely the *phenomenon* of revelation of reality—that is, the phenomenon of reality being revealed to consciousness—that constitutes the designatum of the appellation 'truth,' and this is also the reason for characterizing truth as essentially epistemically qualified. What makes it a phenomenon is the rendering of reality *as such* (that is, reality hitherto unknown) into *known* reality. Reality is revealed to consciousness, and truth (about reality) consists in that revelation. Truth therefore can only be (epistemically) 'revealed' truth. Truth is 'in the picture' only because reality is revealed to consciousness.

The metaphor of 'rendering' used in the preceding paragraph is meant to signify the *intelligibility* of reality. Reality known is 'reality made intelligible' to the knowing subject in terms of the most reliable standards of comprehensibility. But the notion of 'known reality' implies the plausibility of the notion of 'unknown reality.' Unlike truth, which can only be known truth, reality can be either known or unknown. Unknown reality is no less reality for not being known. This is the sense in which it is meaningful to talk about reality which may be beyond all possible knowledge of cognitive subjects with finite intelligence. Being truth consists in its being *attained* by an intelligent subject, and tracking reality by means of intelligence constitutes that attainment. Thus, truth is what it is by virtue of its being the intellectual 'measure' of reality, whereas reality does not have to measure up to anything. Truth may be described as the crucial 'point of contact' between the epistemic subject on the one hand and objective reality on the other.

When finite intelligence cannot measure up to reality, there is no light of knowledge to reveal reality and no 'truth-tracking' as a measure of reality. So the idea of 'reality beyond measure' in respect of the capacity of finite intelligibility is very much available to us, and it is this idea that underpins metaphysical realism, i.e. the theory of objective reality that is mind-independently real. If truth is also objective, its objectivity is derived from that of reality inasmuch as truth is operationalized as the demonstrated ability of finite intelligence to measure up to objective reality. The ontic import of truth is similarly derivative from the fact of its being the measure of reality. At the same time, its epistemic qualification is owed to the intelligence of the epistemic subject involved in attaining truth by demonstrably measuring up to objective reality.

The picture of truth that I have painted presents itself as structurally two-dimensional. It partakes of the ontic dimension in so far as reality provides the objective correlate necessary for truth. Indeed, the identity of truth is its 'tracking' relation with reality. On the other hand, truth is internally tied to the epistemic dimension in so far as it is never in the picture unless and until the intelligent subject attains it in a cognitive encounter with reality. Whereas we grasp the notion of truth from the objective perspective by alluding to the ontic (or reality) dimension, that grasp is incomplete without our realization, from the subjective perspective, that attaining truth consists in the cognitive state's attaining objective adequacy. In other words, the knowing subject has to attain cognitive adequacy of the level that measures up to objective reality. Thus, the subjective and the objective perspectives converge in the complex unity of the concept of truth.

It would be clear now how truth and reality are to be distinguished in respect of the feature of objectivity. Both truth and reality are objective; but the dimension of objectivity is the sole constitutive feature of reality. Reality is qualified as *objective* reality, and the qualification, though redundant, is significant nonetheless. The mention of the redundant qualifier proves to be efficacious in emphasizing that there is no subjective face to reality. In other words, reality is what it *is*, objectively speaking, and it cannot at the same time *be* how it might *appear* to some cognitive subject.

By contrast, truth is Janus-faced: it is objective but has its subjective face as well. And the most important point to draw from this difference is that it is the subjective face of truth that surfaces prominently when we talk about our *consciousness* of reality. For, our consciousness of reality is the *intimation of truth*, provided that we are not under the spell of delusory consciousness.

Reality and intelligibility

Is the relation between the real and the intelligible one of equivalence, so that whatever is real is *ipso facto* intelligible? Is this relation symmetrical? It seems to me that the relation is asymmetrical. Whatever is intelligible is of course real, just as whatever is digestible is edible. But it is not obvious that whatever is real is amenable to intelligibility, where 'intelligible' implies conceptualization by epistemic categories and frameworks on the part of actual and possible minds. What is problematic is whether it is plausible to make the synthetic *a priori* claim that there is more to reality than what meets any actual or possible minds. Is it coherent to form the idea of a reality which is inaccessible to our, or any, consciousness, because none of us or any mind may have the epistemic framework for conceptualization of such a reality?

There are philosophical traditions (classical Indian thought for one) in which the idea of reality which is beyond rational intelligibility in any epistemic framework is considered to be quite coherent. I think the argument for coherence here is that ultimate reality is *noumenal* but comprehensible to supra-rational intuition. Reality in this sense is not an object of conceptual intelligibility, but this is no reason for its incomprehensibility. There is more to comprehensibility than conceptual intelligibility.

Supra-rational intuitive comprehensibility of reality is predicated on a consciousness that affords an intimation of truth of a quite different order and quality. This consciousness is radically distinct from the consciousness that sustains the epistemic subject or mind involved in, and concerned with, conceptual intelligibility of reality. What is most important about this distinction is that the consciousness of intuitive comprehensibility of reality is regarded, in the tradition in question, to be indistinguishable from, hence metaphysically identical with, reality itself. In Vedantic thought, for instance, the ultimate reality, *Brahman*, is the reality of consciousness. And acquiring knowledge of ultimate reality or *Brahman*-consciousness is one with *being* or becoming that very consciousness.

The idea of reality beyond any conceptual intelligibility is no less prominent in the Western tradition, especially in modern Western philosophy represented by Immanuel Kant. Noumenal reality for Kant is an indeterminate 'something out there' entirely beyond epistemic conceptualization. Intelligible reality is *phenomenal*, and there is only the intimation of phenomenal truth, which includes scientific truth.

There is, in Kant, no allusion to a consciousness capable of a non-conceptual intuitive access to noumenal reality-as-such. But the *concept* of noumenal reality figures significantly in his philosophical discourse, despite categorical admission of the impossibility of conceptual intelligibility of noumenal reality. Does this fact suggest that the notion of reality-as-such is nonetheless intelligible? Such a suggestion would be puzzling and paradoxical. How can we have an intelligible concept of reality which is by definition beyond conceptual intelligibility *for* the subject forming that concept?

Of course the puzzling ghost is exorcised if it is clarified that the concept of noumenon is a *limiting* concept – a concept that serves as a heuristic device to indicate the outermost limits of thought or intelligibility. This would be intellectual humility and wisdom of human finitude. But such a charitable interpretation could not be consistent with Kantian philosophy. For the concept of noumenon plays an explanatory role in so far as noumenal reality is held to be the ultimate *causal basis* of the phenomenal reality available to human consciousness. If it is an explanatory concept, it could not be construed as an abstract limiting concept, i.e. a concept of mere abstract intelligibility.

Granted that the concept of noumenon cannot plausibly be construed as a limiting concept, I want to argue, in defence of Kant, that conceptual reason does provide us with a dim intimation of reality-as-such. Neither supra-rationally intuitive nor empirically sensuous, this intimation is *synthetic a priori* rational intimation of the primordial *something* (rather than nothing) – i.e. *being itself* prior to any conceptual determination and epistemic intelligibility. Despite being most indeterminate, the concept of reality as primordial being is intelligible as a pre-conceptual and pre-epistemic condition for the possibility of determinate knowledge.

Reality of consciousness: a self-intimating truth

Two notions of reality can be extracted from what has been discussed thus far. There is the pre-epistemic notion of reality *as such* as an indeterminate 'something' understood in some undefined sense of primordial being. Then there is the notion of reality as the conceptually determined object-domain of epistemic subjects. In the latter case, it makes fairly good sense to talk about the intimation of objective (scientific) truth about reality. But in the former case there cannot be said to be any intimation of truth about reality because it is not an object-domain for any epistemic subject. It is reality only in the sense of being a theoretical posit – a 'something' that is an unconceptualizable presupposition of any conceptualized representation of facts, including scientific representations of the universe.

I now want to claim that the *objective* intimation of reality is not the only sense in which we can talk about having intimation of truth about reality. This is because reality is not just objective reality to which the epistemic subject stands in the mind-to-world relational structure. If truth about objective reality is revealed to subjective consciousness, it also means that there is the reality of subjective consciousness itself to which this truth is revealed. Subjective consciousness is real in so far as it is the source of intelligibility and the means of revelation.

It is intriguing to talk about the intimation of truth about the reality of consciousness. If objective reality is revealed to subjective consciousness, what is the reality of consciousness itself revealed *to*? If the only source of any revelation or intimation is subjective consciousness, then it is clear that consciousness must be revealed to itself. The reality of consciousness must be self-intimating. If revelation of reality is in virtue of consciousness, then revelation itself is a proof of the reality of consciousness. And since revelation is a subjective notion, the dimension of subjectivity is constitutively integral to the reality of consciousness. Consciousness poses itself as a *sui generis* reality because of its constitutive subjectivity, which makes it difficult to understand how consciousness fits into a unified theory of objective reality.

The peculiarity of consciousness as a reality is that it is revealed to itself. Hence it is necessary, if not sufficient, that the existence or reality of consciousness is certified by consciousness itself. This reflexivity or self-intimating character of consciousness has extremely significant epistemological and metaphysical implications. If consciousness adopts an epistemic stance towards itself, there would seem to be the epistemic relation of consciousness-as-subject with consciousness-as-object within the orbit of a *single* consciousness. But a self-intimating consciousness is by nature such that its intimation of itself is not the intimation of an *object*. Since consciousness is constitutively subjective, it can only have subjective intimation of itself – that is, intimation of subject-consciousness of itself *as* subject-consciousness. On the metaphysical front, the necessity of subjective intimation of the reality of consciousness radically problematizes its reality-*status* from the objective perspective. We seem to be forced into admitting the objective existence of something that can exist only in its irreducibly subjective constitution.

In virtue of its essential subjectivity, consciousness makes available the phenomenon of 'lived reality' or 'life-world' (*lebenswelt*), which is a complex web of existence structured by experience. Existence as lived or experiential reality – *life* in short – inherits the property of reflexivity of (self-intimating) consciousness. Since consciousness is woven into the fabric of lived or experiential reality, the truth in that reality would have a self-referential property. It would be a truth about the living subjects themselves, and an intimation of this truth would thus be *self*-intimation.

Truth about life-world or lived reality is grasped in terms that are intentional or meaning constituted and teleological or purposive. But truth concerning life is subject to critical interrogation because meaning and purpose are open to interpretation. Inasmuch as humans are 'self-interpreting animals,'[3] the truth about their lived reality would be largely constituted by interpretative projections of their beliefs and attitudes. Truths about lived reality are therefore partially constituted by the living subjects themselves.

Reality in dispute: between Einstein and Tagore

I have spoken of three kinds of reality: 1) indeterminate reality-in-itself, 2) conceptualized reality as an object-domain, and 3) lived reality. It is at this juncture that the famous debate between Albert Einstein and Rabindranath Tagore can be brought to bear on the distinctions I am making here.[4]

It is not clear whether Einstein ever believed in the first kind of reality and supra-rational intuitive intimation of the truth about it. Most probably he didn't. In the Tagore-Einstein conversation alluded to in the above, he expressed a scientific realist view which he distinguished from the human-centric view of reality. In responding to a remark made by Tagore, he talked about 'two different conceptions about the nature of the universe – the world as a unity dependent on humanity, and the world as a reality independent of the human factor.' But there is no implication in this remark that the possibility of accessing the truth of this 'human-independent' reality ultimately depends on any kind of supra-rational intuitive intimation of it.

It is also not clear whether Einstein would have accepted a revised version of Kantian noumenal reality, construed as an indeterminate primordial being of which we have a kind of speculative inferential grasp as the ultimate basis of determinate knowledge about reality. I can only surmise that he would not have found this reformulated version of noumenal reality nonsensical. For such a reality is consistent with his notion of reality as mind-independently objective. I suppose that Tagore would have found such a notion of reality far too abstract and detached to be a 'human conception' of reality, which he regarded to be the only intelligible conception we can afford to entertain.

On the concept of reality of the second kind, their disagreement with one another needs to be seen in the light of the distinction I have drawn between truth and reality. They both seem to use these two terms interchangeably, although they assign different meanings to them. In my view truth is two-dimensional in that it admits of both epistemic and ontic characterization. Since the epistemic dimension is that of the epistemic subject's *attainment* of truth, truth has a subjective face, given that epistemic subjectivity measures up to ontic objectivity (i.e. reality). So Tagore's insistent denial of the validity of the notion of truth as observer-independent is broadly compatible with my view, so long as the observer is the epistemic subject whose cognitive representation has the adequacy of measuring up to reality.

In contrast to truth, reality is uniquely characterized by objectivity in the sense that being revealed to consciousness is not a constraint on, or condition of, being real. While there can be truth about reality and epistemic consciousness may attain that truth, attaining (or failing to attain) truth about reality has no conceptual bearing whatsoever to reality. Reality is metaphysically independent in the sense that it does not have to measure up to anything, let alone to the human mind, in order to be reality. In fact, it is quite conceivable that reality is ultimately under-determined by any cognitive conceptualization afforded by finite intelligence of any species. It is precisely this liability to cognitive under-determination of reality by any finite intelligent mind of any species that Kant had alluded to in the positing of the concept of noumenal reality.

It is with regard to the so-called noumenal reality that we may wonder whether we can meaningfully talk about truth. We can certainly talk about phenomenal truth, or truth about phenomenal reality in the Kantian sense. Must we also grant that there is noumenal truth analogous to phenomenal truth? One could say, of course, that where there is reality, there is truth about that reality. Granting noumenal reality would then *ipso facto* amount to granting noumenal truth. Truth is, in this sense, a cognate of reality. It thus looks compelling to associate truth with reality in that the former seems to be entailed by the latter.

However, my discussion of the nature of truth would seem to conflict with the claim that the notion of truth is entailed by that of reality. If truth is (epistemically) *revealed* truth in the sense of reality being revealed to cognitive consciousness, then there cannot be any truth about a putative reality which is beyond the cognitive boundary of any finite consciousness. Since Kantian noumenal reality is inaccessible to any finite cognitive consciousness, it is a reality never to be revealed to the consciousness of any species. Therefore, we cannot plausibly talk about noumenal truth inasmuch as there is no possibility of ever coming to know what possibly could be the putative 'truth' about noumenal reality. Strictly speaking, Kantian noumenal reality challenges intelligibility because it sets the limits of intelligibility.

In regard to reality of the third kind, *lived reality*, I suppose Einstein would agree with Tagore, but that would be an agreement *by default*. Since lived reality is essentially experiential, it is by definition consciousness-dependent. But Einstein would add, I believe, that the legitimate criterion of lived reality (namely, consciousness-dependence) is being illegitimately extended by Tagore to apply to reality in general. Lived reality is a small sub-set of reality in general, and what characterizes the sub-set, i.e. mind-dependence, need not characterize the larger set.

Tagore's illegitimate extension is, however, interestingly predicated on the intriguing notion of a 'central personality' or a 'logical person in the centre,' which is presumably the universal counterpart of the 'peripheral' personality of individual consciousness, around which distinct life-worlds of lived reality are constituted. His remark deserves to be quoted here:

> I must emphasize this fact that this world is a real world only in its relation to a central personality. When that centre is taken away, it then falls into pieces, becomes a heap of abstractions, matter and force, logical symbols, and even those,—the thinnest semblances of reality,—would vanish into absolute nothingness, if the logical person in the centre, to whom they are related in some harmony of reason, were nowhere.[5]

Much as personal life-worlds are centred in individual selves in the domain of 'local' reality of human creatures, reality as a whole, or 'global' reality, must also be anchored in a centre, which is the universal personality of a 'logical person' capable of comprehending the larger picture of reality. Subjective anchorage in this 'central personality' is a *logical* requirement of the intelligibility of any metaphysical conception of reality. There is no sense in the idea of a reality entirely detached from what Tagore calls 'the entanglement of the world of self.' The 'logical person' is a metaphysical extrapolation of the empirical individual self, and the extrapolation is necessary if the ultimate picture of reality is not to become incomprehensibly discontinuous with reality intelligible to the empirical self.

Perspectival priority and perspectival reality

There are two theoretical perspectives that vie with one another in this debate on the nature of reality. One accords priority to the perspective of consciousness and self, and reckons talk about reality to be intelligible only with reference to an idealized self, or what Tagore calls 'the logical person in the centre' of reality. The other perspective is to develop a conception of what Thomas Nagel calls a 'centreless universe' – a conception of reality detached from any entanglement of consciousness and the first-person point of view, but inclusive of the self with its first-person perspective on the world.[6] Tagore and Einstein are the respective incarnations of these two opposing theoretical viewpoints – the centred and the centreless viewpoints (as we may call them) on what is reality.

The centred view keeps us at ease because we tend to feel that the intelligibility of reality must have its source in the cognitive self, howsoever universally the cognitive potentialities of this self are conceived. The centreless view might come under critical review because of the uneasy feeling that there is no particular cognitive perspective from which the so-called centreless universe would be intelligible. Indeed, the centreless view is an extrapolation from a *theoretically* posited centre-point, a kind of absolute Archimedean standpoint, from which a perspective-neutral picture of reality is formed.

However, intuition seems to suggest that the centreless 'view from nowhere' may be the only regulative ideal which can prevent our conception of reality from lapsing into species-specific parochialism. The ease of comprehensibility afforded by the centred view of reality may conceal our species-specific cognitive complacency. We may take the view entertained from the human perspective to be representative of the best possible view of the world. And our intended objective view of the world may surreptitiously incorporate a view which privileges the human cognitive subject in the domain of possible cognition.

Against this metaphysical fixation of a worldview that unconsciously accords cognitive centrality to the human self, the so-called view from nowhere de-centres the self and its privileged cognitive depiction of reality. What this view intends to achieve is to make us recognize, and come to terms with, the compelling nature of the perennial question of metaphysics, namely, 'what the world is like, not from the viewpoint of human consciousness, but as a world that includes this viewpoint as one of its contents.'[7] However, the centreless conception of reality is bound to remain interminably incomplete, given that there is a limit to how much we can extrapolate this conception from the most objective view we have from the human perspective. If the threshold of comprehension even of the most abstract understanding of reality is unavoidably conditioned by species-specific cognitive limitations, then all we can afford is a very speculative and sketchy picture of a centreless world.

It is compelling to admit that an objective conception of reality must be formulated in a manner that maximally transcends species-specific peculiarities of perceptual perspectives. But this apparently compelling admission also fails to discern a hidden complexity in the relation between reality and perspective. The objective conception of reality—the conception that reality is objective in the sense of *not* being relative to any perspective—regards any perspectival fact as a contingent and eliminable feature, belonging to what is categorized as *appearance*. It is objective reality over subjective appearance that a true conception of reality is supposed to settle for in the epistemic quest for truth. Any perspectivally characterized fact is apparently a fact but not really a fact at all. The so-called realm of appearance is a pseudo-ontological plane in which any such perspectival fact is located, only to be ultimately discarded as a kind of *persona non grata* in the metaphysical adjudication of the status of reality.

A clear example of the elimination of perspectival facts from the objective picture of reality is provided by the history of modern science in Galileo's theory of properties of matter. This theory is philosophically articulated in John Locke's celebrated distinction between primary and secondary qualities.[8] In the mechanical, atomistic, corpuscularian view of physics that Locke subscribes to following Robert Boyle, the fundamental units of the world are indivisible particles or atoms, which have properties such as solidity, shape, and extension, motion and rest. In this atomistic, mechanistic worldview, there is no place for sensible qualities such as colours, smells, tastes, sounds, warmth and cold and so on. But then, where are these latter qualities of ordinary perception and common bodily feeling located? These qualities, called secondary qualities, are relegated out of objective reality into the realm of the mind or the inner recesses of subjective experience. Thus, the primary/secondary quality distinction is intended to help accommodate these sensible and felt qualities within an atomistic, mechanistic conception of matter.

One may view this distinction as stemming from the thought, as Barry Stroud puts it, that 'the world presented to human beings in their daily lives diverges from, and appears to be a much richer place than, that austere reality described by the atomists.'[9] It is here that we can talk about the world of *appearance* as a unique feature of reality, provided the presence of these appearances is not reducible to atomistic or mechanical properties of the world of matter. What is in question is the metaphysical characterization of appearances, whether we can legitimately talk about the *reality* of appearances, and thereby save appearances from being dismissed as unreal.

Colours, sounds, smells, tastes, cold and warmth are the *ways* the world appear to us as we experience the causal impact of the world upon us. If our *consciousness* of the world is as real as the world we are conscious of, and our consciousness of the world cannot be a fact without its experiential contents consisting of various sensible and felt qualities, then it follows that these felt contents – *appearances* in the experiential plane of the mind – are a fact of the world. In fact, the ineluctable reality of such appearances is a direct consequence of the nature of the reality of consciousness. I cannot be a conscious creature without being conscious *of* the surrounding world. And consciousness of the world entails having the world appear in a certain way in conscious experience. Hence the reality of appearances occurring to consciousness is necessarily integral to, or an intrinsic part of, the reality of consciousness.

However, the reality of appearances is subjective and thus perspectival. What appears as coloured or noisy or sour or foul-smelling or warm and so on is real only to *me*, who is the subject of these experiential states. These are real only to the first-person perspective from which I have any experience of the world. Consciousness with its structurally inbuilt first-person perspective registers the array of appearances. And what it registers figures as real phenomena solely to the subject of that consciousness. These phenomena are not real from the third-person perspective, or from the impersonal objective standpoint. It is for this reason that John Searle refers to the subjective ontology of the mental when he says that 'the ontology of the mental is an irreducibly first-person ontology.'[10] Thus the fact of appearance is a first-personal perspectival fact.

Since appearances are real from the first-person perspective of the conscious subject, the truth about appearances is also a perspectival truth. The reality of appearances consists in their distinctive phenomenological content, that is how they subjectively feel to the conscious subject. For example, the phenomenological content of my visual experience of a yellow mango is *what it is like*, for me, to have the experience. So the truth about this phenomenal reality is *this* specific what-it-is-likeness. It is a subjective, phenomenal truth that is revealed exclusively to the conscious subject who is visually exposed to the mango in broad daylight.

There is no problem or question of acquiring epistemic or cognitive adequacy for the revelation of phenomenal truth of such perceptual appearances. Indeed, this adequacy is already built into the constitution of consciousness and the nature of experience. To have the visual experience of yellowness is *already* to be acquainted with the what-it-is-like content of the visual appearance of yellowness. Thus the yellow-appearance's phenomenal reality is self-intimating or self-revealing. Of course there is a set of objective facts about visual experience of yellowness the knowledge of which depends on acquiring epistemic adequacy. This set of facts would include neuro-physiological details about the visual system of the observer and physical information about the frequency or wavelength of light and the properties of the surface of the mango. But these objective facts are required in order to know the *conditions* under which a yellow visual experience can normally be had by a conscious subject. The conditions are not essential to the truth about the subjective what-it-is-like content of the experience. And to know what really is a visual experience of yellow is to know what it is *like* to have the experience. That knowledge, or access to the intrinsic truth of the experience, is guaranteed to the subject by the very fact of *having* the experience.

It is not just perceptual experiences and bodily feelings of pain and pleasure that constitute the phenomenologically real, perspectival world of the conscious subject. Indeed, the entire experience of lived reality, the life-world of the subject in its manifold entirety, is often a diffuse experience of how life as a whole *appears* to that subject. This 'appearance' of one's own life as one lives it is real to the subject in the robust sense of reality. However, the lived reality of life is ontologically limited to the subject's consciousness of the inner flow of that life. It is real only from the inside, as it were, in so far as the subject feels the life to be one of happiness or unhappiness, smooth or complicated, hope or despair, or a combination of the two opposite qualities in varying proportions of mixture. At any rate, the lived reality of life is real from the first-person perspective of the conscious subject whose life it is. The truth about the reality of my particular life-world is not just revealed to me; it is also an irreducibly personal truth in that no other subject has direct access to the phenomenological reality of my life. Other subjects can at best have an inferential access to the robust reality of my life. Your empathy with my lived reality actually means that you are at your imaginative best in this regard.

What can now be said about the whole topic of reality and intelligibility in a summarized way is this: that reality is both objective and subjective, and the intelligibility of reality is accordingly perspective-neutral or perspectival. Of course, this distinction would not have been necessary were it not for the fact of consciousness. Consciousness is at once necessary for the revelation of reality of either kind. But its role in accessing the truth about perspectival reality, especially lived reality, is peculiarly unique. Unlike the reality of objective existence, the subjectively charged lived reality of life can possibly be changed qualitatively by the power of the conscious subject living that life. If it appears unhappy or despairing, consciousness can exercise its inner power to transform that life into a happy or hopeful one. Appearance can thus be consciously manipulated more or less radically, whereby lived reality is made to undergo a qualitative transformation in its reality status. The intelligibility of lived reality is also an insight into morally transformative self-liberating consciousness.

Notes

1 In *Descartes: The Project of Pure Enquiry*, Bernard Williams provides a compelling description of objectivity in science, which he characterizes as the 'absolute conception' of reality. According to this conception, while there possibly are different human perspectives or representations of the world, the world itself is an independently existing reality. Although different representational perspectives are inevitably resorted to by different human theorists, they can nonetheless recognize their perspectival limitations or distortions in picturing the independent world. The aim of objectivity in science is the collective and cooperative human endeavour methodologically to approach, as closely as possible, the absolute conception of reality – a conception of what there is *anyway* – independent of the human point of view.

2 Perhaps this point about reality not being a matter of status that has to be earned or can be lost is interestingly connected with the famous philosophical dictum, made famous by Emmanuel Kant in his legendary criticism of St Anselm's Ontological Argument for the existence of God, namely the remark that existence is not a predicate or property that can be predicated of any being. According to the Ontological Argument, God necessarily exists because a God without the property of existence would not be 'the being greater than which there is no other being' (that is, the greatest conceivable being). In other words, the argument says that God, the greatest conceivable being, is by definition such a being that it cannot lack existence. Existence is part of the concept of God by ontological necessity. Kant's refutation of this argument is that it wrongly counts existence as another property along with other usual properties like being powerful, knowledgeable, wise etc.

3 That humans are 'self-interpreting animals' – that the self is 'a being who exists only in self-interpretation' – is famously argued by Charles Taylor in Charles Taylor, *Philosophical Papers, Volume I: Human Agency and Language*, p. 3.

4 The historic conversation between Albert Einstein and Rabindranath Tagore on the nature of reality was first published in the *New York Times*, August 10, 1930 by Dmitri Marianoff under the title 'Einstein and Tagore Plumb the Truth – Scientist and Poet Exchange Thoughts on the Possibility of its Existence without Relation to Humanity.' This conversation has most recently been published in Partha Ghose (ed), *Einstein, Tagore and the Nature of Reality*, Appendix I, pp. 221–4.

5 Rabindranath Tagore, *Personality: Lectures Delivered in America*, p. 98.

6 Thomas Nagel, *The View from Nowhere*.

7 Nagel writes that 'if the general human perspective is then placed in placed in the same position as part of the world, the point of view from which this is done must be far more abstract, so it requires that we find within ourselves the capacity to view the world in some sense as very different creatures also might view it when abstracting from the specifics of their type of perspective. The pursuit of objectivity requires the cultivation of a rather austere universal objective self. While we can't free it entirely of infection with a particular human view and a particular historical stage, it represents a direction of possible development toward a universal conception and away from a parochial one' (*op. cit.*, p. 63).

8 John Locke, *An Essay Concerning Human Understanding*, Book II, chapter viii.
9 Barry Stroud, *The Quest for Reality*, p. 7.
10 John Searle, *The Rediscovery of the Mind*, p. 95.

References

Ghose, Partha (ed). 2017. *Einstein, Tagore and the Nature of Reality*, New York, Milton Park, and Abingdon: Routledge.

Locke, John. 1975. *An Essay Concerning Human Understanding*, ed. Peter Nidditch, Oxford: Clarendon Press.

Nagel, Thomas. 1986. *The View from Nowhere*, New York: Oxford University Press.

Searle, John. 1992. *The Rediscovery of the Mind*, Cambridge, MA: MIT Press.

Stroud, Barry. 2000. *The Quest for Reality: Subjectivism and the Metaphysics of Colour*, Oxford: Oxford University Press.

Tagore, Rabindranath. 1917. *Personality: Lectures Delivered in America*, London: Macmillan Company.

Taylor, Charles. 1985. *Philosophical Papers, Volume I: Human Agency and Language*, New York: Cambridge University Press.

Williams, Bernard. 1978. *Descartes: The Project of Pure Enquiry*, Harmondsworth, Middlesex: Penguin Books.

11

THE CREATIVE MIND

A mirror or a component of reality?

Ashish Lahiri

Science, religion, philosophy, poetry

At the early age of 12, after his *upanayana* (the sacred thread ceremony, a rite reserved for Brahmins), Rabindranath's father Debendranath Tagore ritualistically introduced him to the *gāyatrī mantra*.[1] 'The splendid cadence and intonation of this Vedic verse appealed strongly to his feeling for rhythm and his sense of the mysterious sublime. The *gāyatri* remained his lifelong companion and he continued to find in it a source of contemplative insight and strength long after he had discarded the sacred thread.'[2]

However, thanks to the rigorous coaching of his house tutors, he had also learnt that there are certain laws in accordance with which nature works and which are 'testable' (to use a phrase from Karl Popper). That was how the seed of the swing between the sense of the mysterious sublime and that of the comprehensible universe was sown in his mind; wonder being the common factor linking them.

Except in the very early, made-to-order *Brahma-saṃgīts* (which he later hated to accept as his 'own' other than as musical exercises), one can clearly discern the tension between these two approaches to nature – to reality – in many of Rabindranath's creations, particularly in the songs brought together by the poet himself under the categories *Pūjā* (Worship) and *Vicitra* (The Diverse). On the one hand is an unknowable, numinous, *a priori* entity permeating everything that exists; on the other is the play of a set of universal natural laws to be discovered gradually by the human mind through sense data. One can legitimately claim that if the *Upanishads*, among other things, had formed Rabindranath's mind, Rabindranath in turn had edited the *Upanishads* to suit his science-honed perception of reality. What he loved to call the contradiction – or maybe the absence of it – between the Infinite and the Finite becomes better intelligible as a manifestation of this dichotomy. In the process of understanding this dichotomy, one increasingly becomes aware of a restive, open-ended mind torn between many tensions, a mind for which raising the sincere question was often more important than finding a ready answer. It is this awareness that brings Tagore close to the anxiety-prone confrontational psyche of modern humans: 'a modernity which values change more than stability, for which a sense of Becoming matters more than Being.'[3]

Lest one should think that all this pertains to the realm of contemplative philosophy, I must point out that he was emphatic in positing that modern culture should imbibe science – theoretical science – as an inalienable part of its *aesthetics*. Tagore was fully aware of the difference between pursuing science *per se* and deriving aesthetic pleasure out of it. Commenting on his friendship with the scientist Jagadis Chandra Bose, he wrote at the mature age of 77 that what had attracted him to the latter 'was more his (J. C. Bose's) magical instinct than the probing of his reason which startled out secrets of nature before sudden flashes of imagination. In this I felt our mutual affinity but at the same time our difference . . . my own world of visions had their value not in their absolute probability, but in their *significance of delightfulness*.' And yet, in the next sentence he adds, 'All the same, I believe that a part of my nature is logical which not only makes playthings of facts, but seeks *pleasure in the analytical view of objective reality*.'[4] These, then, were the two major components of Tagore's mind vis-à-vis science: 'making playthings of facts' fished out by scientists and deriving 'pleasure' from the 'analytical view of objective reality.'

In the introduction to his book on modern physics and cosmology, *Viśvaparicay* (An Account of the Universe, 1937), dedicated to the physicist Satyendranath Bose (of Boson fame), Tagore wrote:[5]

> Right from childhood, my craving for the taste (*rasa*) of science knew no bounds. I believe I was nine or ten at the time, when Mr. Sitanath Datta [Ghosh] would sometimes drop in on a Sunday. I now know that his stock was not big, but when he explained a simple theory or two of science with illustration, my mind would fill with wonder. I still remember that he would put water on heat, and pouring sawdust into it showed how the hot water, being lighter, streamed upwards and the cold water, being heavier, moved downwards, and that was how the entire mass of water heated up. I still remember the thrill of discovering a continuous mass of water dividing into two streams at the same time.

This is a typical example of seeking 'pleasure in the analytical view of objective reality.' What stirred him was the realization that the heart of something apparently so quiet contained such tremendous turbulence. He was convinced of the restricted nature of the one-dimensional world of our sense experience, and could feel that in order to get at the true nature of things, you have to destabilize them and change the 'given' state. Thus scepticism and not blind acquiescence with passive reality, is the key to knowing nature. You have to confront and not merely worship nature as the giver of all that is, as the *gāyatrī* would prompt you to do.

Einstein and King Pravāhaṇa

For contrast, one might recall a similar, but far more intense encounter of Einstein's who 'experienced a miracle . . . as a child of four or five when my father showed me a compass.' Quoting Einstein's sister, his biographer Abraham Pais informs us that 'It excited the boy so much that "he trembled and grew cold. . . . There had to be something behind objects that lay deeply hidden." . . .' The interesting point, however, is that for the rest of his life Einstein believed that 'the development of [our] world of thought is in a certain sense a flight away from the miraculous.'[6] The phenomena only *appeared* to be miraculous, in reality they were not. They were wonderful, even awe-inspiring, but not miraculous. Miracles were merely reflections of one's ignorance about the myriad laws of nature or of one's commitment to the irrational (that was Spinoza speaking!). Like Rabindranath, Einstein too was drawn to institutional religion in his childhood, but at the age of 12, with his first formal introduction to organized science, he for ever gave up that path. In later life what he eulogized as 'Cosmic Religion' had nothing to do with religion *per se*. Steven Weinberg rightly points out that 'Einstein's religion was so vague that I suspect that he meant this metaphorically . . .'[7] He literally carved out his mind in the immortal quote: 'Subtle is the Lord, but malicious he is not.' The 'Lord,' i.e. Nature, would not reveal His mystery at one go; but eventually reveal He will, in small hiccups, in an exasperatingly tantalizing process. Einstein was a realist *par excellence*. He believed in the real existence of the world and the complete 'knowability' of the world, however tough the task.

Not so with Tagore. Philosophically he was a dualist, not a realist. For the greater part of his life, he had been a believer in religion. He was not a reductionist trying to understand nature in terms of its basic constituents; he was a holistic synthesizer trying to fathom the meaning of the phenomenal universe:

> In dark night the desolate man goes crying
> in frenzy, unveil, oh unveil the mystery.

For him there was always something transcendental, something beyond the reach of human experience. It is the illimitable, he said, that continually spurs on physical science to search for new frontiers. He would fondly refer to King Pravāhana of *Chāndogya Upanishad* (1:8:8)[8] to emphasize the point:

> If man would ever stoop to agree that there was an absolute limit, then his physical science would have dropped anchor and stopped moving long ago. Scientists had once claimed that they had peeled off all the layers of the fundamental elements of the physical world: so absolute was the pealing that the elements were not analysable any further. But see what has happened to that claim. Seated in the heart of things, King Pravāhaṇa goes on propelling all human enquiries further and further beyond every limit.[9]

One can readily see where the two great minds converge and where they diverge. More on that later. For the time being, let us continue with the story of the development of scientific wonder in young Rabindranath.

Of sawdust and stars

A few years after he had learnt the theory of convection of heat from Sitanath Ghosh, he had accompanied his father

> to Dalhousie in the Himalayas. . . . He (Debendranath) would acquaint me with the stars and the planets. He would go further and tell me about the planets' orbital distances from the sun and the duration of each planetary year. I used to take notes. Based on those I penned an immature article. I wrote because I had savoured the taste – this was my first-ever biggish article and its content was scientific information.[10]

'Thrill' and 'taste' are the keywords at this stage. What did the boy taste? What thrilled him? It is easy to reconstruct that he had sensed an isometric unity between the circulating stream of sawdust in the beaker of hot water and the celestial objects moving about in empty space. He was awestruck by the fact that the sawdust, the water, the fire of the burner and the bright objects millions of light years away should obey the same set of physical laws. This is not philosophy. This is something one feels in one's bones, something that could stir one into poetry, into music. How strange that these insignificant, puny, helpless creatures that call themselves *homo sapiens* should be tied to the immeasurably vast universe through the bond of a *single*set of interconnected laws! And science is all about finding out and mastering those laws. This appreciation itself is a wonder, that itself gives a taste of the infinite and the taste of infinity gives one the assurance of a cosmic shelter. What is significant is that Rabindranath got a first glimpse of that sanctuary through science at a very early age, if not before, simultaneously with his initiation to *gāyatrī*.

Later he would explain the emergence of life in these terms: 'Then came a time when life was brought into the arena in the tiniest little monocycle of a cell.' He went on to elucidate how that tiny cell of life could add value to existence by 'overcoming the *obstructive inertia of Nature by obeying Nature's law*.'[11] You have only to break the obstructive inertia of nature, for instance by heating water, and then you come to know about the laws of nature which you cannot but obey, being a part of nature yourself. It is not for you that nature works; it has its own ways. It is for you to discover those hidden, often camouflaged laws, and that is where the whole humanity depends on science. It is this emphasis on the laws of nature, to be discovered by following the scientific method that distinguishes Tagore from most of the other idealistic artists and thinkers of modern India.

On the strictly philosophical plane too, Tagore conceptualizes the holistic integration working through the apparently separate principles of nature. This integration he calls 'a comradeship of creative co-ordination with the entire physical universe.' Here is an interesting example.

> Take, for instance, a piece of coal. When we pursue the fact of it to its ultimate composition, substance, which seemingly is the most stable element in it, vanishes in centres of revolving forces. These . . . can further be analysed into a certain number of protons and electrons. Yet these electrical facts are what they are, not in their detachment, but in their interrelationship, and though possibly some day they themselves may be further analysed, nevertheless the pervasive truth of interrelation which is manifested in them will remain.[12]

In a sense, it would not be an exaggeration to say that throughout his life, through all his creations, Rabindranath has given expression to this one wonder of comradeship, this one taste in many words, many tunes, many hues and many metres.

> In immense space, in immense time, in an immense world,
> I roam – Man, alone in wonder.

How majestic and how irresistible is the use of the word 'wonder' here! The lone human's search for a cosmic shelter, which is at once a 'creative comradeship,' finds expression in this apparently mystic craving:

> In infinite space and time, in these countless effulgent worlds,
> Your gaze is on me – my gaze is on You.

The rebel atom and the *Natarāj*

Another compelling image is that of the 'rebel atom,' which seamlessly fuses with the mythical figure of the dancing *Natarāj*, the creator cum destroyer. In one of the most famous songs of Tagore composed in 1927 we find:

> Your dance renders beautiful the rebellious atoms.
> The light-bells resonate round your ankles, the sun and the moon.
> The inert world comes alive at your dance's life-breaths.
> From eon to eon, in tunes in beats,
> in joy in sorrow, in waves rides this bliss of yours.

Interestingly, while discussing the problem of emergence of life from inorganic matter in the light of the Second Law of Thermodynamics,[13] the famous scientist Ilya Prigogine also stumbled upon that very sculpture – the *Natarāj*, calling that incomparable attitude 'a junction between stillness and motion.'

The fact that the atom and the sub-atomic particles are ever restless, and yet that ferment is incomprehensible to our common sense, never failed to amaze Rabindranath. That scientists could reconcile, explain and formulate such grossly contradictory notions is, for him, one of the greatest triumphs of *homo sapiens*, something that sets the species apart from animals. Talking of the celebrated particle-wave duality of light, he remarks,

> These contradictory notions lie beyond the realm of the common-sense language of human beings. Yet, human beings

> did not fight shy of diving deep into the sea of the incomprehensible. They did not even hesitate to call this stonewall an unceasing dance of electrical particles! They did not fear that they might have gone mad. . . . In fact, science has proved that human beings *are* a race of animals bemused in madness. Whatever they think they perceive is in reality the exact opposite of what it actually is. Animals are never so libellous about themselves.[14]

Once again, we find, more than the facts and the theories of science, he values the wonder that flows from it. This gives him an easy route away from philosophy to creative art.

He characterizes the primeval dance of the atom as 'rebellious' and 'beautiful.' That life-breathing dance sensitizes the 'inert world' into consciousness, causes absolute bliss to undulate in rhythms and tunes, through joy and sorrow. Not stability, nor unwavering stillness, but instability and rebelliousness are at the bottom of all creation.

> In all molecules of universe-body pulsates the shadow of this dance.

Dipanakar Chattopadhyay, a physicist himself, interprets it in the following way:

> Clearly, one step of the dance of *Natarāj* brings forth order and pattern, the other disorder and disharmony. The point in between the two phases Prigogine calls "a junction between stillness and motion". . . . The basic point about *Natarāj*'s dance is that creation involves a combination of many contradictions. Greek mythology talks of the emergence of cosmos from chaos. In our country these two poles have been treated not sequentially, not one after the other, but simultaneously. It is through the conflict between the two that this world assumes myriad forms.'[15]

The moment Tagore uses such images as 'the rebel atom' and envisions the sun and the moon as the resonant 'light-bells' orbiting 'round your ankles,' and finds that the shadow of Śiva's dance pulsates through all the molecules of the body of the universe, the traditional idea of the 'primal rhythmic energy' acquires a new relevance. Philosophy and myth here transmute into poetry through the catalytic agency of modern science. This is what one would like to call the process of poetic internalization of scientific ideas, of which Tagore is a supreme instance.

A whole genre of Tagore songs is devoted to catching glimpses of this dialectical relationship between the two aspects of the same reality: stillness and motion.

Cosmic scepticism sublimated into poetry

As an example of Tagore's cosmic scepticism sublimated into high poetry, one might consider the famous song *Vānī mor nāhi*, composed nearly at the end of his life (1939).

> Words have I none. I can but spread out my heart all over the sky. I am the light-lorn darkling night, gazing with its myriad stars in futile hope at the endless path ahead.
>
> The magic of your flute sounds afar. The tunes come pouring through the silent deeps, floating on the raptured breeze across the slumbrous ocean.
>
> I can but give back echoes of your own strains. Do they ever touch the shore of your dreams beyond the great deep?[16]

For the interpretation that follows, I rely upon Professor Gouri Prasad Ghosh, the translator.[17] 'From across the boundless world of phenomena innumerable waves of stimuli come floating and touch the poet's consciousness all the time. Yet, the poet [i.e., the human mind] has not the capacity to [decisively] respond to the great, mesmerizing stream of information from infinity. All through his life, he has only learnt how to receive these strange and mysterious waves of messages from the universe. As a result of his many-splendoured lifelong pursuit he has now emerged as a sensitive radio receiver which does not yet have the power to transmit counter-signals.' Ghosh goes on to ask: '. . . but is there any message to transmit at all? The profoundest questions of consciousness are sealed behind an impregnable wall of unanswerability.'[18]

Indeed, there is no such thing as the final answer. That is why knowing never ceases. Human beings shall never be able to say, we have learnt all. The histories of science and philosophy are replete with such arrogant fallacies.

Yet, amidst this dimness of the consciousness, there are moments when the innermost door to the poet's soul suddenly opens. 'In such moments he feels as if the great artist who is the source of all these emanations has started playing his flute from a distance beyond calculation.' The poor poet can only produce an echo to that tune. He cannot respond by creating a tune of his own. But who knows the course of even that echo? Does it ever reach its original source? In other words, can a human being ever have the coveted cosmic shelter?

Epistemologically, this is almost the same question raised in the famous 'creation hymn' of the*Rigveda*:

> Who really knows? Who will here proclaim it? Whence was it produced?
>
> Whence is this creation? The gods came afterwards, with the creation of the universe. Who then knows whence it has arisen?
>
> Whence this creation has arisen – perhaps it has formed itself, or perhaps it did not – the one who looks down on it, in the highest heaven, only he knows – or perhaps he does not know.[19]

However, we are not concerned here with epistemology. What concerns us is the process of sublimation through which this epistemological scepticism turns into the pure poetry of feeling. Plainly, there are two possible ways: one, to submit oneself completely to a transcendental being, as Tagore has done in numerous poems and songs; and two, to internalize that scepticism into an aesthetic experience – as Tagore has done in this specific instance. As Ghosh says, 'Tagore has written quite a few amazing poems of spiritual faith. He is unique in giving artistic shape to faith internalized as intense feeling. However, even deep faith sometimes has its elements of unconscious self-betrayal, which may strike an identical chord of weakness in the reader's or the listener's mind.' That is why spiritual creations are often so moving. However, this particular song is anything but spiritual. Here scepticism reigns supreme. Ghosh writes: 'In *Vānī mor nāhi*, a feeling of truth of the highest order has attained artistic form – a feeling unadorned with the cosmetic touches of wish-fulfilment.'

Two years before composing this song, Tagore had written *Viśvaparicay*. There he had explained in scientific terms the great significance of *homo sapiens*' scientific quest – a quest that is as unrelenting as it is ever incomplete. The dreams of attaining a final theory on the one hand, and the insignificance of that attempt in the context of the immense complexity of the universe, in other words, the 'provisionality' (to borrow a term from J.D. Bernal) of all scientific theory – both these are there. Tagore shows a finely balanced appreciation of the two aspects.

Three stages

The general intellectual realization and the creative internalization of scientific scepticism are processed through three distinct stages. At the first stage he is a diligent, if capricious, student of science – one for whom science and poetry belong to two great but different worlds, mutually exclusive. He sees and wields modern science as the weapon for decimating the gross and unreasoning Hindu revivalism that was raising its head in his early youth. Strangely, at the turn of the twentieth century he himself was to an extent influenced by such hidebound ideas. However, as has happened so many times in his life, he quickly came out of this somewhat claustrophobic phase. The second decade of the twentieth century saw a new Tagore. Typical of this new phase was his delightful essay *Āmār Jagat* (My World), written in 1914, shortly after he had returned from a momentous tour of Europe and the USA.

Here he records a mock debate between a poet and his scientist friend. He apparently accepts the superior position of reductionist science so far as knowing objective truths is concerned. He does not question the validity of such knowledge. One finds the patronizing scientist almost patting the poet on the back. Presently, however, the poet hits back, questioning the scientist about his methodology. When describing the apparently fixed stars, astronomy says, they only appear to be fixed, because you are looking at them from such a great distance. If you look at them from a sufficiently close distance, or can devise some mathematical substitute for such close observation, you will find that they are actually moving fast. Very fine; then how about the earth? The surface beneath your feet is obviously flat; then why do you say that the earth is actually round? Here your point is that if you looked at the earth from a distance, you would get a real picture of its true shape. Why these double standards? Why do you discount your immediate sense data in one case and accept them in the other? Heaping instance upon instance, drawn from the fields of physics, botany and psychology, the poet himself arrives at the synthesis that both these viewpoints are necessary: he calls them the *near side* and the *far side* of the same thing. One may characterize these as the reductionist and the holistic approach respectively. In his typical *Upanishad*-honed fashion, he calls them the finite and the infinite. A complete mind, he asserts, has place for both. Indeed, only that mind is complete which can simultaneously comprehend the invisible ever-restless sub-atomic particles and an apparently solid nugget of iron, or the extensively porous leaf when looked under a microscope and the tightly shaped leaf seen by the naked eye. He generalizes that the concepts of time, space, position, distance, nearness, movement, stillness, inside, outside – all have this dichotomy. He joyously quotes the famous lines from *Īśā Upanishad* to drive his point home: *Tadejati tannaijati taddūre tadvantike* (It moves, it moves not; it is far, it is near). For him, the mind is not merely a mirror that reflects reality, but a component (*upakarana*) of that reality.

But then, he asks, if each mind is a separate component of reality, then what about objectivity? Strangely, in answering this question, the poet takes recourse to artistic experience itself. A particular song composed in the rāga *miyan-ki-malhar* expresses very exclusive feelings, those that might pertain to a particular day's rain experienced by a particular composer at a particular time and space. And yet, that composition touches the heart of millions at different times and spaces. How could that be possible without something common to the millions of individual minds that were at the same time *components* of the feelings generated by the physical reality of rain?

From here, it is only a step to conclude that the poet need not feel shy when his feelings are apparently at variance with the findings of the scientist. Art has a world of its own which is simply out of bounds for reductionist science. The point to note is that he is not against reductionist science *per se*, as long as it remains confined to the domain of science. Reductionist science, he seems to say, has no business to meddle in the affairs of art. Thus, in the end the patronizing scientist gets a much-deserved rebuff from his poet friend.

The essay *Āmār Jagat* (My World, 1914) thus represents a watershed in Tagore's ideas so far as the problems of subjectivity of art and the so-called objectivity of reductionist science is concerned. From this time onwards, we find in him a finer appreciation of the philosophical problems of science, as well as its relationship with the other creative aspects of man. At the culmination of this stage, he makes a surprising statement in 1932:

> If you ask me what pure modernism is, I would say it's looking at the world not with a subjective personal attachment, but with an objective clinical detachment. That is what constitutes a pure vision; that detached vision is bliss. Modern science analyzes reality with a detached mind; modern poetry should also do the same, for that is what is eternally modern.[20]

It is 'surprising,' because earlier he had accused modern (i.e. reductionist) science of intruding upon literature and thereby robbing it of its essential beauty, which, he explained, resided in purely arbitrary personal attachment. Literature, he said, was characterized essentially by its prejudices and caprices, which was in direct opposition to the impersonal objectivity of rational science. He cites a beautiful example from his own *Citrāngadā*. Arjuna is indulging in voyeurism, when he stealthily looks at young narcissistic Citrāngadā undressing before bathing in a pond, rapt at the exquisiteness of her own form. Arjuna is excited and overwhelmed. This, Rabindranath appears to be saying, can be treated in two ways. One is the strictly Freudian interpretation, which is perfectly in order as science. However, the moment that interpretation interferes with and dominates over individual artistic presentation, it kills art. Tagore felt that Western literature of the modern period had fallen victim to just such an aggression from science. The extreme obsession of literature with the purely physical aspects of sex, which he saw as a 'nuisance,' was a manifestation of this breakdown.

Therefore, when he jumps from this position to its opposite and says that 'poetry should also do the same, for that is what is eternally modern,' one is a little taken aback. However, he goes further. He says:

> The mathematician no doubt engrosses himself in the profound symmetry permeating high-level mathematics, in the unity of forms. The fact of its orderliness is not only epistemic, it belongs also to the sphere of deep feeling: there you get pure bliss. It finds expression at the apex of knowledge where it is free of any utilitarian concern. There knowledge attains liberty. One naturally wonders why this has not been the subject of poetry. The reason, of course, is that its experience is esoteric, its access denied to the common person.

Thus, the poet confidently says, there is no intrinsic impediment to high mathematics being the subject of poetry! What a change from the position that science was poking its ugly nose in the delicate affairs of poetry!

Tagore and Heisenberg: holism, scepticism and uncertainty

This is perhaps the place to recount the tête-à-tête between Tagore and Werner Heisenberg (1901–1971), the physicist of 'Uncertainty Principle' fame.

For an account of the Tagore-Heisenberg talk one relies on the diary of D. M. Bose, the well-known Kolkata physicist. He writes that Sommerfeld[21] and Heisenberg visited India in 1928,

> both of whom stopped for a few days in Calcutta on their way to attend another International Conference in Japan. Heisenberg appeared one day without any previous introduction in the University College of Science. Some of us . . . arranged a lunch at Firpo's for Heisenberg. Rabindranath was in Calcutta at that time. Heisenberg having expressed a desire to see him, it was arranged that we were to take him the same afternoon to Jorasanko (the poet's residence). On arrival, we found that [the poet's son] had arranged a fine tea for us. We left Heisenberg to have a talk with the poet. I do not remember what was the substance of his talk, but Heisenberg was very much impressed by the poet's illuminating personality which reminded him of a prophet of the old days.

We get two diametrically opposite accounts of what transpired between the mature poet (67 at that time) and the young scientist (then only 27). In his book *Uncommon Wisdom*, Fritjof Capra writes, 'In 1929 (actually 1928) Heisenberg spent some time in India as the guest of the celebrated Indian poet Rabindranath Tagore, with whom he had long conversations about science and Indian philosophy. This introduction to Indian thought brought Heisenberg great comfort, he told me. He began to see that the recognition of relativity, interconnectedness and impermanence as fundamental aspects of physical reality, which had been so difficult for himself and his fellow physicists, was the very basis of Indian spiritual traditions. "After these conversations with Tagore," he said, "some of the ideas that had seemed so crazy suddenly made much more sense. That was a great help for me." '[22]

We have seen that already in 1914 Tagore was quoting from *Īsha Upanishad* (*Tadejati tannaïjati taddūré tadvantiké*) to assert that concepts like position, distance, nearness, movement, stillness, inside and outside were not absolute. One can presume that Tagore showed that there indeed was a philosophical strand of thought that went well with Heisenberg's mathematical findings about physical reality, even though it appeared to contradict our 'common-sense,' i.e., Newtonian, notions.

But let us not be too sure about the impression produced by Tagore on the young Heisenberg, as claimed by Capra. Krishna Datta and Andrew Robinson write: 'However, according to Heisenberg's wife (who was not a scientist), "my husband was not too much impressed by his [RT's] thoughts. The mixture of eastern and western philosophy in his thoughts did not really convince him" (Elisabeth Heisenberg to Authors, 3 Oct. 1990). Since Heisenberg himself wrote nothing on the meeting that has survived, it is hard to know his true impressions.'[23]

Read between the lines, D. M. Bose's comment would also tend to support Elisabeth Heisenberg: 'I do not remember what was the substance of his talk, but Heisenberg was very much impressed by the poet's illuminating personality which reminded him of a prophet of the old days.'[24] Had Heisenberg been *scientifically* impressed by Tagore's philosophy, is it not reasonable to assume that he would certainly have said something about it to a fellow physicist, namely D. M. Bose? Instead, what he said had been standard fare for all Westerners: a prophet-like, even Christ-like personality, whatever it might mean.

However, it clearly shows that Tagore read and interpreted the *Upanishads* in his own non-conformist manner. His familiarity with modern science reinforced that non-conformism. His interpretation and understanding of the *Upanishads* became increasingly individualistic and increasingly tinged with scepticism.

Nor is this a mere unfounded conjecture. There is solid evidence that Tagore regarded the *Upanishads* not as revealed texts, but as having their base in the real temporal life of the people. In 1924, he wrote, '. . . the words contained in them had their full context in the life of the people of that period, who spoke them. Divested of that vital atmosphere, a large part of the language of these great texts offers to us merely its philological structure and not life's subtle gestures which express through suggestion all that is ineffable.'He was interested more in discovering those 'suggestions' than in theological reasoning: 'the *Upanishads* are based not upon theological reasoning, but on experience of spiritual life. And life is not dogmatic; in it opposing forces are reconciled – ideas of non-dualism and dualism, the finite and the Infinite, do not exclude each other.' As for the nature of the subtle 'suggestions' permeating the *Upanishads*, he clearly says that 'Suggestion has its unanalyzable code which finds its depth of explanation in the living hearts of the people who use it. . . . All poetry is full of such words, and therefore poems of one language can never be properly translated into other languages. . . .'[25] He quotes from Keats's 'Ode to the Nightingale' to drive home the point that mere philological correspondence with the English words will not make for a happy Bengali rendering. This analogy with poetry makes it clear that here he is advocating a creative reading of the *Upanishads*. That is why he was able to express his own idiosyncratic, *non-theological* and sceptical understanding of reality on the one hand, and at the same time communicate with the new ideas of modern science. Heisenberg's experience is a case in point.

In a remarkable passage, Rudolph Otto, the well-known Christian theologian who wrote the *Idea of the Holy*, comments on Tagore's spiritual yet non-theological perception of reality. Tagore 'wanted neither to expound any theological theory before us nor to lecture upon any condensed doctrine of Indian mode of belief, but to lay open his . . . totally personal perception which by no means can be extricated from this person . . .' Narrating how Tagore explained to him the special significance of *Īśā Upanishad*, Otto very shrewdly noted 'how in reality an age-old religious perception here renews itself in a modern soul . . . how an ancient heritage at the same time changes into a contemporary one. . . .'[26]

It is this non-theological and non-dogmatic view on reality that sets Tagore apart from most of the other idealist thinkers of modern India. One would like to characterize this approach as 'cosmic scepticism.'

Tagore-Einstein dialogue: the dualist and the realist dilemmas

This cosmic scepticism constituted the core of Tagore's famous dialogue with Einstein in 1930 in Berlin. To any percipient reader it should be clear that throughout the talk, Tagore is eloquent and expansive, while Einstein is rather abrupt, often unresponsive, speaking more in staccato phrases than in well-connected sentences. It is noteworthy that during the conversation Dmitri Marianoff, who was to be Einstein's son-in-law, was present and he observed: '. . . it seemed to an observer as though two planets were engaged in a chat.'[27] Einstein himself said in a letter that 'My conversation with Tagore was rather unsuccessful because of difficulties of communication and should, of course never have been published.'[28] Why was Einstein so hesitant and unhappy about publishing the conversation, which contrasts strangely with the eagerness with which Tagore and his friends published it?

Dipankar Chattopadhyay has examined this issue in some detail. He notes that on the request of Romain Rolland, Einstein agreed to write a contribution to *The Golden Book of Tagore* in 1931. 'Initially it was decided that he would write on the duties of intellectuals against war. However, Einstein changed the subject absolutely and instead wrote an article titled *About Free Will*. . . . This article is tinged with some irony, which might be a result of Einstein's dissatisfaction with the conversation.' Einstein's article begins with the observation:

> If the moon, in the act of completing its eternal way round the earth, were gifted with self-consciousness, it would feel thoroughly convinced that it would travel its way of its own accord on the strength of a resolution taken once for all.[29]

Indeed this difference had been the axis around which the 'conversation' had spiralled: human consciousness and natural laws. There we find Tagore repeatedly emphasizing the classical Berkeleyan stand that sense data had no 'real' significance and that reality existed only in the consciousness of what he called 'Universal Man.' Einstein was fully familiar with this syllogism – he was well versed in Berkeley. However, in 1930, when this conversation took place, he had become too much of a 'realist' to accept that. Heisenberg recounts that by 1926 Einstein 'thought that every theory in fact contains unobservable quantities. The principle of employing only observable quantities simply cannot be consistently carried out.' This was patently opposed to Mach's philosophy to which Einstein had earlier subscribed. When Heisenberg pointed this out, Einstein answered simply, 'perhaps I did use such philosophy earlier, and also wrote it, but it is nonsense all the same.'[30]

That change of stand also lay at the root of Einstein's discomfort with quantum mechanics which depicted a probabilistic as against a deterministic picture of nature. He was too great a mathematician and physicist not to agree with Heisenberg's principle; but philosophically he was an absolute realist, which meant he believed that the senses could, in principle, deliver us a true – not a probabilistic – picture of reality, provided we had the right theories and the right methodology. Nevertheless, he confessed that this was only a belief. Thus, we find the curious phenomenon that an idealist poet was arguing in terms of *logic* and a realist scientist was confessing that 'This is my *belief*, although I know well that it is not fully demonstrable.'

What is the essence of that belief? In his article in *The Golden Book of Tagore*, Einstein says: 'Man defends himself from being regarded as an impotent object in the course of the universe. But should the lawfulness of happenings, such as unveils itself more or less clearly in inorganic nature, cease to function in front of the activities in our brain?' He points outs that 'alcohol and other sharply controllable factors' do have an influence 'on our thoughts, feelings, and activities.' Therefore, 'determinism does not stop before the mystery of human will.'[31]

This was an obvious retort to Tagore's stand that 'the infinite personality of man comprehends the universe. There cannot be anything that cannot be subsumed by the human personality, and this proves that the truth of the universe is human truth.' Einstein is ready to concede that concepts like 'beauty' are human-specific, but scientific 'truth' is not. Tagore explains beauty as 'the ideal of perfect harmony, which is the universal being.' One becomes increasingly aware of Tagore's unconcern for scientific methodology, about what constitutes 'scientific truth.' He does not seem to appreciate that as a working scientist Einstein's primary duty was to be true to his method. The scientist cannot follow the dictates of his mind, unless these were commensurate with his mathematics and his method. That is why he could not deny the worth of quantum mechanics, even as he was strongly opposed to it philosophically. Not so with Tagore. He was not answerable to scientific rigour. He was free to follow the dictates of his creative soul. Thus, he could easily say: 'Truth must be essentially human.'

Tagore, however, does not dismiss the difference between scientific truth and what he hypothetically calls the 'universal truth.' On the contrary, he agrees that there are two kinds of truth: one scientific, the other transcendental. He acknowledges that scientific truth 'can only be reached through the process of logic,' but that very logic is only 'human'; hence scientific truth is dependent on human beings. Scientific truth is merely a human truth, nothing beyond that. The obvious corollary is that science, human-dependent as it is, cannot lead us to an objective understanding of reality. No wonder he ends up denying the comprehensibility of the universe by invoking *māyā*: 'what appears to be true to the human mind, and therefore is human, may be called *Māyā*, or illusion.'[32]

This may be very subtle philosophy, but for a working scientist, this must be stupefying. In Einstein's own words in a different context, 'Why should anybody go to the trouble of gazing at the stars if he did not believe that the stars were really there?' One can almost see why Einstein thought that this conversation should not have been published. For the two were talking at cross-purposes: one was talking about how to account for the real universe of which man was only a part; the other was talking about 'merging the individual in its infinity.' The 'two planets' were indeed chatting from their individual orbits!

Notes

1 'We meditate on the adorable glory of the radiant sun; may he inspire our intelligence.' Trans. S. Radhakrishnan, *The Principal Upanishads*, New Delhi: Indus, 1994, p. 299.

2 Krishna Kripalani, *Tagore: A Life*, New Delhi: National Book Trust, India, 2001, pp. 19–20.
3 Sankha Ghosh, *Nirman ar Srishti* [Construction and Creation], Kolkata: Visva Bharati, 1982, p. 298. Translation mine.
4 Sisir Kumar Das (ed), 'Jagadis Chandra Bose', in *The English Writings of Rabindranath Tagore*, vol. III, New Delhi: Sahitya Akademi, 1996, p. 826.
5 *Viśvaparicay* [An Account of the Universe, 1937], *Rabindra Rachanavali*, Centenary Edn. vol. 14, Kolkata: West Bengal Government, 1961, p. 822.
6 Abraham Pais, *Subtle is the Lord*, Oxford: Oxford University Press, 1997, p. 37.
7 Steven Weinberg, *Dreams of a Final Theory*, London: Vintage, 1993, p. 194. Also, regarding Einstein's famously recorded belief in 'Spinoza's God', see p. 205: 'And it does not seem to me to be helpful to identify the laws of nature as Einstein did with some sort of remote and disinterested God. The more we refine our understanding of God to make the concept plausible, the more it seems pointless.'
8 'What is the goal [i.e. substratum or basis or final principle] of the yonder world?' He replied, 'One should not lead beyond this world-support. We establish the *Sāman* on the world as support for the *Sāman* is praised as the support.' (1:8:7). Then Pravāhaṇa, son of Jīvala, said to him, 'Verily, indeed, your *Sāman*, O son of Śālāvat, has an end. If someone now were to say, 'Your head will fall off,' surely your head would fall off.' (1:8:8) Radhakrishnan comments: 'The enormity of the error is suggested by the statement that your head will fall off if one utters a curse like that.' See Radhakrishnan, *The Principal Upanishads*, p. 352. It is indeed remarkable that while a Brahmin school of thought emphasized 'on the world as support for the *Sāman*', King Pravāhaṇa, a *kshatriya*, should consider this empirical idea as an enormous and limiting error, so much so that 'your head will fall off if one utters a curse like that.' The same Tagore-Einstein controversy, as it were, was being carried on here under a different set of terms of reference.
9 Rabindranath Tagore, *Mānusher Dharma* [Man's Religion, 1932], *Rabindra Rachanavali*, Kolkata 1961, vol. 12, p. 575. Translation mine. Though of a similar title, the Bengali work is not the same as the English *The Religion of Man.*
10 *Viśvaparicay*, *RR*, vol. 14, p. 822.
11 Rabindranath Tagore, 1931. *The Religion of Man*, New Delhi: Indus, 1994, p. 9.
12 Ibid., p. 14. One may note in passing the anticipation of the later 'electrical facts' like quarks in this passage.
13 'Heat cannot be transferred by any continuous self-sustaining process from a colder to a hotter body.' E. B. Uvarov and Alan Isaacs, *The Penguin Dictionary of Science*, London: Penguin Books, 1993, p. 444.
14 *Mānusher Dharma*, *Rabindra Rachanavali*, vol. 12, p. 576. Translation mine.
15 Dipankar Chattopadhyay, *Rabindranath o Vijnan* [Tagore and Science], Kolkata: Ananda Publishers, 2000, pp. 213–14. Translation mine.
16 Gouri Prasad Ghosh, *Rabindranath Tagore: Thirty Poems and One Hundred and Thirty Songs*, Kolkata: Manabmon and New Delhi: Supernova Publishers & Distributors, 2007, p. 110.

17 Gouri Prasad Ghosh, *Rabindrakavyer Silparup* [The Art of Tagore's Poetry], 2nd ed., Kolkata: Mallik Brothers, 1995, pp. 96–100. Translation mine.
18 Words in parentheses are mine. – A. L.
19 *Rigveda*, X.129.6, trans. Wendy Doniger O'Flaherty in *The Rig Veda: An Anthology*, Harmondsworth: Penguin Books, 1981, pp. 25–6.
20 *RR*, vol. 14, p. 348.
21 Sommerfeld also met Tagore at Santiniketan and had a talk with him on the upper terrace of Udayana. He has left a beautiful account of his impression of the poet.
22 Quoted in Chattopadhyay, *Rabindranath o Vigyan*, p. 252.
23 Krishna Datta and Andrew Robinson, *RabindranathTagore: The Myriad-minded Man*, Kolkata: Rupa, 2000 ed., p. 443.
24 Dipankar Chattopadhyay, *op. cit.*
25 See S. Radhakraishnan, *The Principal Upanishads*, Appendix A, pp. 939–40.
26 Rudolf Otto, 'Preface to The Book "Tagore's Confession" ', in *The Golden Book of Tagore*, Kolkata: Rammohun Library & Free Reading Room, 1990, Addenda, 1931 reprint, p. 17. English translation from the German by Saumendra Palit.
27 Dipankar Chattopadhyay, *op. cit.*, p. 243.
28 Ibid., p. 245.
29 Albert Einstein, 'About Free Will', *The Golden Book of Tagore* (1931), 1990 reprint, p. 12.
30 Ibid.
31 Ibid.
32 Some philosophers have raised the question whether Tagore was right in translating the Sanskrit word *māyā* as 'illusion'. According to them, *māyā* is also a cosmic principle that seeks to explain the existence and creation of the world. I am not competent to judge the point. However, translating the Sanskrit word as 'the veil of illusion', a standard dictionary defines *māyā* in much the same sense as Tagore's: 'The way in which the world is experienced and which disguises the unity of reality.' (Simon Blackburn, *The Oxford Dictionary of Philosophy*, Oxford: Oxford University Press, 1996, p. 235).

References

Blackburn, Simon. 1996. *The Oxford Dictionary of Philosophy*, Oxford: Oxford University Press.

Chattopadhyay, Dipankar. 2000. *Rabindranath o Vijnan* [Tagore and Science], Kolkata: Ananda Publishers.

Das, Sisir Kumar (ed). 1996. Jagadis Chandra Bose. In *The English Writings of Rabindranath Tagore*, vol. III, New Delhi: Sahitya Akademi.

Datta, Krishna and Andrew Robinson (eds).2000. Rabindranath *Tagore: The Myriad-minded Man*, Kolkata: Rupa.

Doniger O'Flaherty, Wendy. 1981. *The Rig Veda: An Anthology*, Harmondsworth: Penguin Books.

Einstein, Albert. 1931. About Free Will. *The Golden Book of Tagore*, 1990 reprint.

Ghosh, Gouri Prasad. 2007. *Rabindranath Tagore: Thirty Poems and One Hundred and Thirty Songs*, Kolkata: Manabmon and New Delhi: Supernova Publishers & Distributors.

———. 1995. *Rabindrakavyer Silparup* [The Art of Tagore's Poetry], 2nd ed, Kolkata: Mallik Brothers.

Ghosh, Sankha. 1982. *Nirman ar Srishti* [Construction and Creation], Kolkata: Visvabharati.

Kripalani, Krishna. 2001. *Tagore: A Life*, New Delhi: National Book Trust, India.

Otto, Rudolf. 1990. Preface to the Book "Tagore's Confession". *The Golden Book of Tagore*, Kolkata: Rammohun Library & Free Reading Room. English translation from the German by Saumendra Palit, 1931 reprint.

Pais, Abraham. 1997. *Subtle is the Lord*, Oxford: Oxford University Press

Radhakrishnan, S. 1994. *The Principal Upanishads*, New Delhi: Indus.

Tagore, Rabindranath. *Mānusher Dharma* [Man's Religion, 1932], 1961. *Rabindra Rachanavali*, Centenary ed., vol. 12, Kolkata: West Bengal Government.

———. 1961. *Viśvaparicay* [An Account of the Universe, 1937], *Rabindra Rachanavali*, Centenary ed., vol. 14, Kolkata: West Bengal Government.

———. 1931. *The Religion of Man*,New Delhi: Indus, 1994.

Uvarov, E. B. and Alan Isaacs. 1993. *The Penguin Dictionary of Science*, London: Penguin Books.

Weinberg, Steven. 1993. *Dreams of a Final Theory*, London: Vintage.

Appendix A

MY MEMORIES OF EINSTEIN

Sir Rabindranath Tagore

I met Einstein during my first visit to Germany after the war. I was then deeply impressed by his great simplicity. There was nothing stiff about him – there was no intellectual aloofness. He seemed to me a man who valued human relationship and he showed toward me a real interest and understanding.

I recall that we talked on that occasion of whether or not I believed in modern industrial improvements to help us in our modern life. I told him then, and have no reason to change my opinion now, that they were essential to our physical well-being, and in as much as nothing could stop these improvements, we should seek to use wisely what man's ingenuity had created out of his necessity. For we had reached that degree of civilization when we could no longer scratch with our fingers; we were using our intelligence to overcome through machinery the weakness of our limbs. Both Einstein and I held similar opinion that it was necessary for us, by such means as mechanical invention offered, to make use of Nature's store-house.

When I again visited Germany in the summer of last, I was invited to come to see Einstein, this time in his own home at Kaputh, a short distance from Berlin. The Professor had a very beautiful retreat there, built high up on a hill. He came down to the road to meet me. His shock of white hair, his burning eyes, his warm manner again impressed me with the human character of this man who dealt abstractly with the laws of geometry and mathematics. Together we walked slowly up the hill, and on his balcony we had afternoon tea. Between us was a sympathy, which only the barriers of language made awkward. His mastery of English is not enough for conversation and I cannot speak German. The interpreter between us must have had a not too easy task.

My mind was full of the matter contained in my Hibbert Lectures, which I had just delivered in Oxford, during May, 1930, and was now putting into book shape under the title of *The Religion of Man*. Einstein's daughter Margaret, who is an artist, came with us to Russia. Later in the summer the Professor and I met again, at the home of Dr. Mendel, who is occupied with the investigation in the field of cancer. Once more I was able, in Einstein, to note the innate modesty of the man. He has a vitally acquisitive mind, and while I haven't much training to meet him on mathematical grounds, we had many discussions together which approached the boundary line of human and abstract conceptions of Reality.

I could readily see that Einstein believed my Universe was limited by human conception, and he was convinced that there was some Truth which was independent of human mind. I claim that the Individual has relationship with the Divine Man, who is within as well as beyond us. Our religion is not cosmic, but it concerns our personal being which lives and has its ideal of good and evil. Science makes no such distinctions of a moral or aesthetic nature, and deals with one factor of existence. Science has nothing to do with personality. On the other hand Religion has nothing to do with mere knowledge of abstract Truth.

Einstein has often been called a lonely man. In so far as mathematical symbols help to liberate the mind from the trivialities, I suppose he is a lonely man. His is what might be called a transcendental materialism, which reaches the frontiers of metaphysics, where there can be utter detachment from the entanglement of the world of self. To me both Science and Arts are expressions of our spiritual nature where they are above our biological necessities, and possessed of an ultimate value.

We talked long and earnestly about my "Religion of Man". Einstein is an excellent interrogator. He punctuated my thoughts with terse remarks of his own, and by his questions I could measure the trend of his own thinking. He asked me to explain fully what seemed to him my "purely human conception of the universe". It is a simple, irresistible conception. The Truth of this world, whether perceived by our senses or conceived by our logic, necessarily belongs to the human mind with all its limitations. Our reason in its harmony with the universal mind has a comprehensive view of things but that view is human as well as our reason and the universal mind has its human character only in perfection.

The Infinity of Humanity is the Perfect Man. Thus the future is merely a prolonged present. Not only is there a physical Universe but there is also moral perfection. Something within us urges us towards the more and more perfect. The tiger is not under this spell, but the savage, when he creates his first piece of art, shows that he is striving to go beyond himself, to extend his limits. This effort is drawn toward a point far away. The tiger, in physical well-being, is quite satisfied. But there is something in Man that is always bringing out his personality, his inner being, always suggesting further adventure. If there was nothing but blind groping, we would never desire to reach the goal of perfection. I see human progress in spite of its innumerable blunders has reached a far higher plane of meaning than what it had at the starting point. There is movement in humanity always going on toward a centre of gravity and this centre is the great Spiritual Man.

I believe this answered Einstein's desire to know whether I held that the Divine was isolated from the world. To me, the great Spiritual Man is a fact to which we belong, a fact not fully fathomed but existent and being realized from within and from without. The highest impulse that we have in our search is not for what we know but for what we do not know.

It is characteristic of the human being that he aspires toward the Unknown. If his search had no meaning for him, the attraction which he feels would lead him to idiocy. The Unknown has led Science, no one can deny that – from the vague to the gradual illumination. While the amoeba turns around in a perpetual repetition, Man's movement is always progressively toward an objective reality. His steadying pivot is his belief that Truth is there, which beckons him even when he does not see it. This is the call to the Individual from the Universal Man – and it is this call that carries us toward the Universal Man. Animals are happy because they hear it not; Man is not happy, yet he is eager for the Adventure.

"So this is the realization of the human entity?" asked Einstein. He grasped that I regarded the Universal Truth of Existence not as an abstraction but as a reality spiritually related to personality. For such Truth can be valued by reason of its contribution to the dignity of our own being, to the realization in us of an inner fulfilment. Here, to my mind, is the call of Religion.

As a Poet I resort to figures of speech. The Infinite Being is like one who plays a flute. Through the music of beauty and love we are brought outside our own self-centered limitations. We sacrifice ourselves because the Central Ideal of Truth calls through these strains of Infinite Being. We become more true in others, we come into greater intimacy with our own Reality in others. There is an eternal spring of Love in the centre of all things. So it is with Truth. We are approaching nearer Central Truth perfected, and that is the meaning of our Evolution. But this Truth has its harmony with human mind, otherwise it would remain absolutely beyond our comprehension. The Truth as we know it or ever may know it, whether physical or metaphysical, is relative to our own knowing mind. I do not say that there may not be a Truth which is beyond our utmost capacity of knowledge but for us it is as good as naught.

I could see that Einstein held fast to the extra-human aspect of Truth. But it is evident to me that in human reason facts assume a unity of Truth which is only possible to a human mind. While Logic belongs to man's mind, I can conceive of another field of logic for another type of mind other than ours. So, when Einstein talked to me of Pythagorean geometry, I could conceive of an existence where Space is not the factor, but where Time reigns as it does in music. It is not impossible that a mind can live in time and have no notion of space. For such a mind, existence might be the kind comprehended in music.

However, the function of religion, according to me, is to bring the individual into harmony in reason, in love indeed with the Supreme Man, the Universal being.

In August 1930, the two of us met again. I had just come from Coblenz, where the Youth Movement was in full blossom. In 1921, I was witness to the pathos of the young Heidelberg students who came to me emaciated from privation. They said, 'We have lost faith in our old teachers'. I felt that this was a glorious time for the young. They were rebellious against the old who had once guided them and had just betrayed them. They wished to build up the future through the fervour that stirred within them – a new future unburdened by the old tradition which had led them to disaster. The young were entering an adventure of building anew, and they moved under an almost hypnotic spell. They desired freedom. This desire was more real to the German youth than to the youth elsewhere. These young people were sensing what I had long felt to be the salvation of civilized man: that we need the primitive life for the sheer vigour of it. With our own hands we must have contact once more with the soil. We must live a life of simplicity – a self-imposed return to nature – but not in the spirit of the Barbarian. Civilization tends to grow blasé, and in Coblenz, I found that only by going back to Mother Earth was youth being won over. The Youth Movement in Germany came to life under the Mother-touch of Nature. In civilization one easily becomes sophisticated, but one does not so easily become wise. So that these young men, who in 1921 were so despondent, had now grown into the joyous realization of the true functions of life. Machinery is not the be-all of their existence. Not that they do not know the uses and the value of Machinery, but they have voluntarily gone back to certain aspects of the primitive humanity in order to find themselves in a renewal of life.

Possibly it was due to my recent illness of that summer that Einstein led me away from abstract matters after a few sallies into them. And so it was we had the following conversation which was taken down by a friend who was present on the occasion.

Appendix B

THE TAGORE–EINSTEIN CONVERSATION ON THE NATURE OF REALITY

Note on the nature of reality

(A conversation between Rabindranath Tagore and Professor Albert Einstein in the afternoon of July 14th, 1930, at the Professor's residence in Kaputh.)

T. You have been busy hunting down with mathematics the two ancient entities, Time and Space, while I have been lecturing in this country on the eternal world of Man, the universe of Reality.

E. Do you believe in the Divine as isolated from the world?

T. Not isolated. The infinite personality of Man comprehends the Universe. There cannot be anything that cannot be subsumed by the human personality, and this proves that the Truth of the Universe is human Truth.

I have taken a scientific fact to explain this – Matter is composed of protons and electrons, with gaps between them, but matter may seem to be solid [without the links in the spaces which unify the individual electrons and protons]. Similarly humanity is composed of individuals, yet they have their interconnection of human relationship, which gives living unity to man's world. The entire universe is linked up with us [as individuals] in a similar manner; it is a human universe. I have pursued this thought through art, literature and the religious consciousness of man.

E. There are two different conceptions about the nature of the universe: (1) The world as a unity dependent on humanity. (2) The world as a reality independent of the human factor.

T. When our universe is in harmony with Man, the eternal, we know it as Truth, we feel it as beauty.

E. This is the purely human conception of the universe.

T. There can be no other conception. This world is a human world – the scientific view of it is also that of the scientific man. [Therefore, the world apart from us does not exist; it is a relative world, depending for its reality upon our consciousness.] There is some standard of reason and enjoyment which gives it Truth, the standard of the Eternal Man whose experiences are through our experiences.

E. This is a realization of the human entity.

T. Yes, one eternal entity. We have to realize it through our emotions and activities. We realized the Supreme Man who has no individual limitations through our limitations. Science is concerned with that which is not confined to individuals; it is the impersonal human world of Truths. Religion realizes these Truths and links them up with our deeper needs; our individual consciousness of Truth gains universal significance. Religion applies values to Truth, and we know this Truth as good through our own harmony with it.

E. Truth, then, or Beauty is not independent of Man?

T. No.

E. If there would be no human beings any more, the Apollo of Belvedere would no longer be beautiful.

T. No.[!]

E. I agree with regard to this conception of Beauty, but not with regard to Truth.

T. Why not? Truth is realized through man.

E. I cannot prove that my conception is right, but that is my religion.

T. Beauty is in the ideal of perfect harmony which is in the Universal Being; Truth the perfect comprehension of the Universal Mind. We individuals approach it through our own mistakes and blunders, through our accumulated experiences, through our illumined consciousness – how, otherwise, can we know Truth?

E. I cannot prove scientifically that Truth must be conceived as a Truth that is valid independent of humanity; but I believe it firmly. I believe, for instance, that the Pythagorean theorem in geometry states something that is approximately true, independent of the existence of man. Anyway, if there is a *reality* independent of man, there is also a Truth relative to this reality; and in the same way the negation of the first engenders a negation of the existence of the latter.

T. Truth, which is one with the Universal Being, must essentially be human, otherwise whatever we individuals realize as true can never be called truth – at least the Truth which is described as scientific and which only can be reached through the process of logic, in other words, by an organ of thoughts which is human. According to Indian Philosophy there is Brahman, the absolute Truth, which cannot be conceived by the isolation of the individual mind or described by words but can only be realized by completely merging the individual in its infinity. But such a Truth cannot belong to Science. The nature of Truth which we are discussing is an appearance – that is to say, what appears to be true to the human mind and therefore is human, and may be called *māyā* or illusion.

E. So according to your conception, which may be the Indian conception, *it is not the illusion of the individual, but of humanity as a whole.*[1]

T. The species also belongs to a unity, to humanity. Therefore the entire human mind realizes Truth; the Indian or the European mind meet in a common realization.

E. The word species is used in German for all human beings, as a matter of fact, even the apes and the frogs would belong to it.

T. In science we go through the discipline of eliminating the personal limitations of our individual minds and thus reach that comprehension of Truth which is in the mind of the Universal Man.

E. The problem begins whether Truth is independent of our consciousness.

T. What we call truth lies in the rational harmony between the subjective and objective aspects of reality, both of which belong to the super-personal man.

E. Even in our everyday life we feel compelled to ascribe a reality independent of man to the objects we use. We do this to connect the experiences of our senses in a reasonable way. For instance, if nobody is in this house, yet that table remains where it is.

T. Yes, it remains outside the individual mind, but not the universal mind. The table which I perceive is perceptible by *the same kind of consciousness which I possess.*[2]

E. If nobody would be in the house the table would exist all the same – but this is already illegitimate from your point of view – because we cannot explain what it means that the table is there, independently of us.

Our natural point of view in regard to the existence of truth apart from humanity cannot be explained or proved, but it is a belief which nobody can lack – *no primitive beings even.*[3] We attribute to Truth a super-human objectivity; it is indispensable for us, this reality which is independent of our existence and our experience and our mind – though we cannot say what it means.

T. Science has proved that the table as a solid object is an appearance and therefore that which the human mind perceives as a table would not exist if that mind were naught. At the same time it must be admitted that the fact that the ultimate physical reality is nothing but a multitude of separate revolving centres of electric force, also belongs to the human mind.

In the apprehension of Truth there is an eternal conflict between the universal human mind and the same mind confined in the individual. The perpetual process of reconciliation is being carried on in our science, philosophy, in our ethics. In any case, if there be any Truth absolutely unrelated to humanity then for us it is absolutely non-existing.

It is not difficult to imagine a mind to which the sequence of things happens not in space but only in time, like the sequence of notes in music. For such a mind such conception of reality is akin to the musical reality in which Pythagorean geometry can have no meaning. There is the reality of paper, infinitely different from the reality of literature. For the kind of mind possessed by the moth which eats that paper literature is absolutely non-existent, yet for Man's mind literature has a greater value of Truth than the paper itself. In a similar manner if there be some Truth which has no sensuous or rational relation to the human mind, it will ever remain as nothing so long as we remain human beings.

E. Then I am more religious than you are!

T. My religion is in the reconciliation of the Super-personal Man, the Universal human spirit, in my own individual being. This has been the subject of my Hibbert Lectures, which I have called "The Religion of Man". In science we go through the discipline of eliminating the personal limitations of our individual minds and thus reach that comprehension of Truth which is in the mind of the Universal Man.

Notes

1 In *The Religion of Man* this phrase appears as 'it is not the illusion of the individual but of humanity as a whole.'

2 In the typescript at the Rabindra-Bhavana Archive and *The New York Times* the phrase appears as 'some kind of consciousness (which) we possess'.

3 In *The New York Times* the phrase appears as 'not even primitive beings.'

Taken from "*The Religion of Man*" (Appendix II) supplemented from the typescripts in the Archive at Rabindra-Bhavana and *The Kenyon Review*, New Series, Vol. 23, No. 2, Cultures of Creativity: The Centennial Celebration of the Nobel Prizes (Spring, 2001), pp. 20–25, published by Kenyon College, which incoprporates the version that appeared in *The New York Times*.

Appendix C

THE TAGORE–EINSTEIN DIALOGUE ON YOUTH, CAUSALITY AND MUSIC

(A conversation between Rabindranath Tagore and Professor Albert Einstein in the afternoon of August 19th, 1930, at the residence of a common friend, Dr. Mendel)

Einstein and Tagore

T. What has impressed me this time in Germany is the work of the German Youth Movement. They are building up a community life on simple and large issues, not by reverting to medievalism but by applying the best gifts of the modern age to the service of a social ideal. Those splendid youths I saw in a Jugendherberge near Coblenz are not only managing their entire organization of lecture-tours, village work, social unions, but even building their own houses, cooking their food, bringing fuel from the woods and personally attending to the minutest detail of daily life without depending on outside help for it. They are full of enthusiasm for the living of life itself, and every function of life therefore is joyous to them, rousing the feeling of adventure and self-expression.

E. The Youth Movement we have in Germany today had, I believe, its origin in England.

T. Yes, may be, but England has nothing like what you have in this country – you are so thoroughly earnest and vital in all that you do here that every new idea finds somehow its full velocity. In England the one institution which delighted me is Woodbroke near Birmingham, where the ideal of community life is maintained through social and educational activities related to our fundamental human aspirations. In Germany, however, as I have said, the vigour and the rich harmony of a complete life finds unique expression.

E. Do these young men near Coblenz permanently live in that colony or is it a centre which they occasionally visit?

T. Many of them live and work there and I believe they also go out in batches on tours in their own country and outside. When once the centre is there, the element of adventure can also find its expression in works done elsewhere. It is this which makes their Youth Movement natural and progressive.

I was discussing with Dr Mendel today the new mathematical discoveries, which tell us that in the realm of infinitesimal atoms chance has its play; the drama of existence is not absolutely predestined in character.

E. The facts that make science tend toward this view do not say good-bye to causality.

T. Maybe not; but it appears that the idea of causality is not in the elements, but that some other force builds up with them an organized universe.

E. One tries to understand how the order is in the higher plane. The order is there where the big elements combine and guide existence, but in the minute elements this order is not perceptible.

T. This duality is in the depths of existence – the two contradictions of free impulse and the directive will which works upon them and evolves an orderly scheme of things.

E. Modern physics would not say they are contradictory. Clouds look as one from a distance, but if you see them near they show themselves as disorderly drops of water.

T. I find its analogy in human psychology. Our passions and desires are unruly, but our character subdues these elements into a harmonious whole. Is it analogous to this in the physical world? Are the elements rebellious, dynamic with individual impulse and is there a principle in the physical world which dominates them and puts them into an orderly organization?

E. Even the elements are not without statistical order; elements of radium will always maintain their specific order now and ever onward just as they have done all along. There is then a statistical order in the elements.

T. Otherwise, the drama of existence would be too desultory – it is the constant harmony of chance and determination which makes it eternally new and living.

E. I believe that whatever we do or live for has its causality; it is good however that we cannot look through it.

T. There is in human affairs an element of elasticity – some freedom within a small range which is for the expression of our personality. It is like the musical system in India which is not rigidly fixed as is the Western music. Our composers give a certain definite outline, a system of melody and rhythmic arrangement and within a certain limit the player can improvise upon it. He must be one with the law of that particular melody, and then he can give spontaneous expression to his musical feeling within the prescribed regulation. We praise the composer for his genius in creating a foundation along with a superstructure of melodies, but we expect from the player his own skill in the creation of variations of melodic flourish and ornamentation.

In creation we follow the central law of existence, but if we do not cut ourselves adrift from it we can have sufficient freedom within the limits of our personality for the fullest self-expression.

E. That is only possible where there is a strong artistic tradition in music to guide the people's minds. In Europe music has come too far away from popular art and popular feeling and has become something like a secret art with conventions and traditions of its own.

T. So you have to be absolutely obedient to this too complicated music. In India the measure of a singer's freedom is in his own creative personality – he can make permutations and combinations of notes according to the law of the melody prescribed but in all this he is guided by his own artistic conscience. He can sing the composer's song as his own if he has the power to creatively assert himself in his interpretation of the general law of the melody which he is given to interpret.

E. It requires a very high standard of art fully to realize the great idea in the original music so that one can make variations upon it – in our country the variations are often prescribed.

T. If in our conduct we can follow the law of goodness, we can have real liberty of self-expression. The principle of conduct is there but the character which makes it true and individual is our own creation. In our music there is a duality of freedom and prescribed order.

E. Are the words also free? I mean to say, is the singer at liberty to add his own words to the melody which he is singing?

T. Yes. In Bengal we have a kind of song – *Kirtan* we call it – which gives freedom to the singer to introduce parenthetical comments as he sings the song. This occasions great enthusiasm as the audience is constantly thrilled by some beautiful spontaneous sentiment freshly added to the song by the singer – phrases which had not been there in the original song itself.

E. Is the metrical form quite severe?

T. Yes, quite. You cannot exceed the limits of versification; in all his variations the singer must keep the rhythm and the time, which is fixed. In European music you have a comparative liberty about time but not about melody, but in India we have freedom of melody with no freedom of time.

E. Can the Indian music be sung without words? Can one understand a song without words?

T. Yes, we have songs with unmeaning words – sounds which just help to act as carriers of the notes. We have two different types of songs in India. In Bengal poems and music combine with each other, in the north-west of India they have often songs with unmeaning or insignificant words. The idea is supplied by the music, the words remaining passive. In Bengal the words and the musical notes meet half-way and continue in the organic creation of a song.

E. In Greece there has been like what you have in Bengal – music with poems.

T. In North India music is an independent art – not the interpretation of words and thoughts. The music is very intricate and subtle and is a complete world of melody by itself.

E. It is not polyphonous?

T. Instruments are used, not for harmony but for keeping time [with] the music and for adding to the volume and depth.

E. European music is not older than 400 years; it was only from then that music, polyphonous music, has been introduced.

T. Has melody suffered in your music by the imposition of harmony?

E. Sometimes it does suffer very much. Sometimes the harmony swallows up the melody altogether.

T. Melody and harmony are like lines and colours in pictures. A simple linear picture may be completely beautiful – the introduction of colour may make it vague and insignificant. Yet colour may by combination with lines create great pictures so long as it does not smother and destroy their value.

E. It is a beautiful comparison; the line is also much older than the colour. It seems that the structure of your melody is much richer than here. Japanese music seems to be so.

T. It is so difficult to analyse the effect of Eastern and Western music on our minds. I feel deeply moved by the Western music – I feel that it is great, that it is vast in its structure and grand in its composition. Our own music touches me more deeply by its fundamental lyrical appeal. European music is epic in character, it has a broad background and is Gothic in its structure.

E. Yes, yes, this is very true.

T. But this music is immense – I can never forget how much I was affected by its power once when a Hungarian lady played on her violin some pieces of music, both classical and modern.

E. When did you first hear European music?

T. At 17 I first came to know it intimately when I first came to Europe, but even before that time I had heard European music in our own household. I had heard the music of Chopin and others at an early age.

E. There is a question which we Europeans cannot properly answer, we are so used to our own music. We want to know whether our own music is a conventional or a fundamental human feeling, whether to feel consonance and dissonance is natural or is it a convention which we accept.

T. Somehow Piano confounds me. The violin pleases me much more.

E. It would be interesting to study the effects of European music on an Indian who has never heard it.

T. Once I asked an English musician to analyse for me some classical music and explain to me what are the elements that make for the beauty of a piece.

E. The difficulty is that the really good music, whether of the East or of the West, cannot be analysed.

T. Yes, and what deeply affects the hearer is beyond himself.

E. The same uncertainty will always be there, about everything fundamental in our experience, in our reaction to art, whether in Europe or in Asia. Even the red flower I see before me on your table may not be the same to you and me.

T. And yet there is always going on the process of reconciliation between them, the individual taste conforming to the universal standard.

Berlin, 19th August 1930.

Taken from the typescript in the archive at Rabindra Bhavana. (Reproduced in *The Kenyon Review*, New Series, Vol. 23, No. 2, Cultures of Creativity: The Centennial Celebration of the Nobel Prizes (Spring, 2001), pp. 25–30, published by Kenyon College, and also in *Asia*, March 1931, p 139.)

INDEX